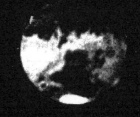

The Planet Mars

William Sheehan

The University of Arizona Press Tucson

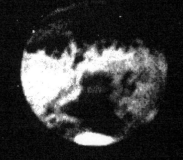

The Planet
Mars

A History of Observation & Discovery

The University of Arizona Press
Copyright © 1996
The Arizona Board of Regents
All rights reserved
Manufactured in the
United States of America

01 00 99 98 97 96
6 5 4 3 2 1

Library of Congress
Cataloging-in-Publication Data
Sheehan, William, 1954–
The planet Mars : a history of observation
and discovery / William Sheehan.
p. cm.
Includes bibliographical references (p.)
and index.
ISBN 0-8165-1640-5 (cloth). —
ISBN 0-8165-1641-3 (paper)
1. Mars. I. Title.
QB641.S484 1996
523.4′3—dc20   96-4485
    CIP

British Cataloguing-in-Publication Data
A catalogue record for this book is available
from the British Library.

TO BRENDAN & RYAN

and the Martians of the next generation

Contents

Figures

Preface

A century ago, at the height of what might be referred to as the "canal furor," Camille Flammarion published the first volume of his great work, *La Planète Mars,* which summarized what was then known about the planet. In his preface he described how he hesitated between two methods of presenting the state of Martian knowledge—in special chapters dealing with topics such as continents, seas, polar caps, and so on; or chronologically, in the order in which the facts had been obtained. He at length decided on the latter approach, "mainly," he wrote, "because it seemed to me to be the more interesting . . . and also because it provides a better account of the gradual development of our knowledge." So it has seemed to me, and I have done likewise. What follows, then, is a history of Martian exploration from the earliest stirrings of human curiosity about the planet right up to the present time when, after a lull of twenty years and after suffering through the disappointments of the Russian *Phobos* and American *Mars Observer* missions, we stand again poised on the verge of a more vigorous phase of exploration of the planet.

I MUST here acknowledge the help I have had in preparing this volume. Dr. Richard McKim, director of the Mars Section of the British Astronomical Association, and Thomas Dobbins were generous in giving of their valuable time to read through the manuscript, and they made many helpful comments. So did two anonymous reviewers for the University of Arizona Press. Dr. Patrick Moore very kindly put at my disposal one of only four existing copies of his English translation of Flammarion's *La Planète Mars,* which has been indispensable, and granted me permission to quote from it and from his earlier translations of the works of E. M. Antoniadi. Drs. Audouin Dollfus and Henri Camichel provided much information about their own studies of Mars as well as about the work of other French pioneers—these were men who received the torch of the great Antoniadi. For help in tracking down material on G. V. Schiaparelli, I am grateful to Signor Luigi Prestinenza; and for permission to use material from the Schiaparelli Archives, my thanks to Professor Guido Chincarini, director of the Osservatorio Astronomico di Brera. Dieter Gerdes, curator of the Heimatverein Lilienthal, was a valuable source of information about J. H. Schroeter. The help of Michael Sims of the Special Collections of Vanderbilt University Archives, Dorothy Schaumberg of the Mary Lea Shane Archives of the Lick Observatory, and Judith Lola Bausch and Richard Dreiser of the Yerkes Observatory is acknowledged. Though my interest in Mars began in my early youth (as has been the case for so many who have been fired with enthusiasm about the planet), my scholarly habits were greatly stimulated by my visit to Lowell Observatory in 1982. I owe much to the encouragement of the late William Graves Hoyt, historian, whose own researches on Percival Lowell resonate in this volume, and Arthur Hoag, director of the Lowell Observatory. Michael J. Crowe, of the University of Notre Dame, also encouraged my research at an early stage. A number of astronomers shared their observations and observing lore related to the red planet, especially Tom Cave, Tom Dobbins, Rodger Gordon, Walter Haas, Harold Hill, and Alan Lenham. Barry Di Gregorio scouted out information regarding the future spacecraft missions. Finally, I am grateful to Richard Baum for his inspiration over many years, and to my wife, Deborah, for helping me in more ways than I can name.

WILLIAM SHEEHAN
Hutchinson, Minnesota
June 1995

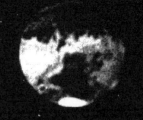

The Planet Mars

Motions of Mars

For thousands of years it was only a blood-red dot among the starry host—a nameless denizen of the trackless night. Sometimes, when it veered closer to the Earth and shone like a burning coal in the darkness, it must have roused terror among primitive sky watchers, only to fade away into relative obscurity and be forgotten once more. By the time the Egyptians settled their civilization along the banks of the Nile, it had become familiar enough to receive a name—Har décher, the Red One. The Babylonians referred to it as Nergal, the Star of Death, and the Greeks too associated it with warfare and bloodshed—it was the Fiery One, or the war god, Ares—one and the same with the Roman god Mars.

The Babylonians made careful astronomical observations and developed a sophisticated system of arithmetical computations for predicting astronomical phenomena such as eclipses. Their purposes were strictly calendrical and religious, however, and they never attempted to explain the reasons for any of the movements they observed. Superstition was widespread, and many astronomical events were regarded as ominous—not just eclipses but even the risings of Venus were viewed as omens.

It is among the early Greeks that we must look for the first stirrings of a more rational perspective. They identified Mars as one of the five "wandering" stars, or planets, which move relative to the "fixed" stars. Two of the planets—Mercury and Venus—always remain close to the Sun in the sky; their distances never exceed 28° and 47°, respectively, and they may pass between the Earth and the

Sun (inferior conjunction) or behind the Sun (superior conjunction). This behavior, as we now know, is due to the fact that their orbits lie inside that of the Earth.

The outer planets—Mars, Jupiter, and Saturn—can appear opposite the Sun in the sky, a situation that is, of course, never possible for a planet that is closer to the Sun than the Earth. When planets appear thus, they are said to be at opposition. It is then that they attain their greatest brilliance. They rise when the Sun sets and set when the Sun rises, so they are highest above the horizon at midnight.

Mars's usual motion among the stars is from east to west. Around the time of opposition, however, it suddenly stops, reverses direction, and moves "retrograde" for a time, then stops again and resumes its usual motion from east to west. (Jupiter and Saturn do this as well, but because they move more slowly—and travel through smaller arcs—the movements are less obvious than in the case of Mars.) So baffling were these motions that Mars was the despair of the naked-eye astronomers. The Roman Pliny, who perished while trying to observe (too closely) an eruption of Vesuvius in A.D. 79, called it *"inobservabile sidus"*; and at least one later astronomer who attempted to calculate the motions of Mars is said to have become deranged in his mind, and in a fit of rage to have bumped his head against the walls![1]

After completing a loop, Mars resumes its westward drift relative to the fixed stars. Its light grows gradually weaker as it approaches and finally passes behind the Sun (into superior conjunction). It then emerges from the Sun into the morning sky and brightens again, until, after two years and two months, it comes once more into opposition and shines like a burning coal upon the night sky.

THE ANCIENT Greeks took for granted that the Earth was the center of the universe. They also assumed that the planets moved uniformly in perfect circles. Unfortunately, uniform motion around simple circles did not account for the complicated movements the Greeks actually observed, and they faced the problem of "saving the phenomena"—showing how the observed movements could be reconciled with their principle of uniform circular motions.

One ingenious scheme was introduced by Eudoxus of Cnidus, a mathematician and contemporary of Plato (indeed, he stayed two months in Athens as a pupil at the Academy, though much of his life was spent in Egypt). In the fourth century B.C., Eudoxus developed his system of homocentric spheres, according to which the observed motion of a planet was produced by the independent motions of several internested spheres, each centered on the Earth. The scheme was able to account for the retrograde movements well enough, but it did not explain why, if the spheres shared a common center, the planets varied in their brightness—in the case of Mars, some fiftyfold.

The most obvious explanation for the fluctuations in brightness was that the Earth was not the center of all the motions. By 250 B.C., Aristarchus of Samos, who at least from our modern perspective was the greatest of the ancient astronomers, worked out a complete heliocentric system in which all the planets circled the Sun. Aristarchus regarded the Earth as an ordinary planet: it rotated on its axis once every twenty-four hours, and it traveled in a circular path around the Sun with a period of one year. With one bold stroke, he had solved the problem of the retrograde movements, which were now seen to be reflections of the Earth's orbital motion—they are apparent displacements as the planets are viewed from different points as the Earth pursues its course around the Sun.

Unfortunately, Aristarchus was too far ahead of his time, and the later Greeks did not follow his lead. His theory did not account exactly for the observed motions, mainly because, as we now know, the orbits of the planets are not exactly circular, as Aristarchus assumed. The most famous Greek astronomers who lived after him, including Apollonius and Hipparchus, returned to the geocentric system.

In their hands the infamous system of epicycles took shape. The system received its greatest elaboration through the efforts of Claudius Ptolemy, who lived at Alexandria in the second century A.D. and worked it out in detail in his great book *The Almagest* (The greatest)—thus, the system of epicycles is more commonly known as the Ptolemaic system. Each planet was taken to move around a small circle known as an epicycle, which in turn moved around a larger circle (the deferent) centered on the Earth. The combined motion of its epicycle and deferent caused each planet to swing in near the Earth at times, thus producing the retrograde movements. Even apart from the retrograde movements, the planets had a variable motion along the Zodiac; thus Ptolemy made their deferents slightly eccentric to the Earth. Even this did not account for the planets' movements, however, so Ptolemy took still greater liberties. He found a point, the *punctum aequans,* from which the planet's motion around the deferent would appear uniform—it was located a short distance from the center in the opposite direction from the Earth (a construction known as the "bisection of the eccentricity"). In practical terms, this involved abandoning the principle of uniform circular motions altogether, but it was brilliantly successful as a device for calculating the observed motions of the planets.

Ptolemy's system, despite its artificiality, remained the last word in astronomy for a thousand years. Although the Dark Ages fell on western Europe, putting an end to astronomical investigations there, astronomy was not completely snuffed out. In the East, in Baghdad, Arab scientists continued to observe the stars, and they also attempted to make minor adjustments in the Ptolemaic theory. Only with the revival of learning in Europe, however, was real progress made. The all-important step was taken by Nicolaus Copernicus, a Polish canon at the Cathedral of Frauenburg. Seventeen centuries after the scheme first occurred to the far-sighted Aristarchus, Copernicus reintroduced the heliocentric system. The Sun rather than the Earth was the center of the system, and once again the fact emerged—especially evident in the case of Mars—that the retrograde movements were mere reflections of the Earth's own motion in its orbit. "This happens," wrote Copernicus, "by reason of the motion, not of the planet, but of the earth changing its position in its great circle. For since the earth moves more rapidly than the planet, the line of sight directed toward the firmament regresses, and the earth more than neutralizes the motion of the planet. . . . The inequality attains its maximum for each planet when the line of sight to the planet is tangent to the circumference of the great circle."[2]

Unfortunately, like Aristarchus's successors Copernicus found that his hypothesis of simple circular orbits around the Sun did not agree exactly with the observed movements of the planets, and he too was forced to introduce complications, including bringing back the cumbersome eccentric circles and epicycles, although in all fairness it must be admitted that he reduced their number. He was also guilty of some surprising inconsistencies. Since the apparent motion of the Sun—actually a reflection of the Earth's orbital motion—is variable during the year, it is convenient to introduce the concept of the mean Sun, whose position is defined by the average rate of the Sun's annual motion. Copernicus referred the planetary motions to the mean Sun rather than to the true Sun, and this forced him to assume that the Earth's orbit had a variable inclination. Despite these shortcomings, he had grasped the essential point; his great book *De Revolutionibus Orbium Caelestium* (On the revolutions of the celestial spheres) is one of the immortal works of science. It appeared in 1543, the year of his death—it is said that the first copies reached him on his deathbed.

Copernicus's work did not find immediate acceptance. The fiercest resistance came from theologians, but many astronomers were also opposed to his views, including Tycho Brahe, the greatest astronomer of the generation that came after Copernicus's death.

Tycho was an entirely different sort of man from Copernicus. Whereas Copernicus was a theorist first and foremost and made only a few of his own observations, Tycho was mainly an observer— one of the greatest who ever lived.[3]

He was born in 1546, three years after Copernicus died. He was adopted at birth by a well-to-do uncle, who destined him for a career in statecraft and sent him at age sixteen to the University of Copenhagen to study law. While there, on August 21, 1560, Tycho witnessed a partial eclipse of the Sun, and it changed the direction of his life. The date of the eclipse had been predicted by astronomers, and Tycho, as his early biographer Pierre Gassendi wrote, "thought of it as divine that men could know the motions of the stars so accurately that they could long before foretell their places and relative positions." Before long, he had obtained a copy of Ptolemy's *Almagest* and worked through it. His uncle disapproved of these studies and sent him away from Denmark, entering him instead at the University of Leipzig with a tutor named Vedel to look after him. Unfortunately, the move was to no avail; Tycho was nothing if not strong-willed. He studied law during the day, and at night, while Vedel slept, stole out to view the stars.

By now Tycho had realized that astronomical tables were not as accurate as he had at first supposed, and also that their reform depended on obtaining more accurate observations. This is what he resolved to do. Not long afterward, his uncle died, and there was no longer anything standing in his way.

From Leipzig, Tycho went to the University of Rostock. During his time there he became engaged in a heated argument with a colleague over a mathematical point. They decided to settle their dispute by fighting a duel, which ended with Tycho having part of his nose cut off (he promptly replaced it with a new one made of copper and wax). From Rostock he went to the University of Basel, then finally returned to Denmark and established a private observatory at Herre Vad, on the estate of another well-to-do uncle.

At Herre Vad, Tycho observed a brilliant new star, or nova, which appeared in the constellation Cassiopeia in 1572. It remained conspicuous for a time, then began to fade, but during the period when it was visible Tycho was able to show that it was exceedingly remote: to all intents and purposes, it was located in the sphere of the "fixed" stars. The book he wrote about the star made him famous, and soon after it was published he received from King Frederik II of Denmark an offer he could not refuse. Frederik granted Tycho the use and revenues of Hven, an island in the Baltic between Elsinore and Copenhagen. Tycho accepted, and in 1576 set up the most splendid observatory in Europe. The instruments there were the best of their day, though needless to say, all were meant to be used for naked-eye observations because the telescope had not yet been invented.

For the next twenty years, Tycho worked at compiling a star catalog whose positions could be trusted to within two or three minutes of arc, and built up an extensive archive of careful observations of the planets, including Mars, which he observed at every opposition beginning with that of 1580. In 1583, he noted that near opposition Mars moved retrograde at a rate of nearly half a degree a day; this proved that Mars could approach much closer to the Earth than the Sun, which was true in the Copernican system but not the Ptolemaic. Nevertheless, Tycho was still not entirely satisfied with the ideas of Copernicus. He adopted a compromise position, known as the Tychonic system, in which the Earth remained at the center; the planets went around the Sun, while the Sun in turn circled the Earth.

It is often said that Tycho was an ill-tempered, quarrelsome man.

Unfortunately, this seems actually to have been the case; and he was also an imperious landlord—he was intensely disliked by nearly everyone on the island, and after Frederik's death the Danish court saw fit to cut off his funds. In 1596, Tycho left Hven, taking along with him his observations as well as the more portable of his instruments. He went first to Germany, and then, at the invitation of the Emperor Rudolph II of the Holy Roman Empire, settled at Prague, in Bohemia. The terrible religious wars between Protestants and Catholics were then in full swing, and in 1600 Tycho was joined by a young Protestant mathematician, Johannes Kepler, who had been expelled from his position as mathematician at Graz, Austria, because his religion differed from that of the Catholic Archduke Ferdinand. At that particular time, Tycho and another assistant, Christian Severinus (or Longomontanus, as he styled himself), were working on the theory of the motion of Mars, and Kepler was assigned to the same monumental task. Kepler later remarked that "had Christian been occupied with some other planet, I would have been started on the same one."[4]

Tycho, who was touchy and jealous of his observations, gave Kepler limited access to his records, and their relationship was undoubtedly strained at times. Miraculously, they managed to avoid a complete break, and on October 24, 1601, Tycho died suddenly, of a bladder ailment. (His last words were, "Let me not seem to have lived in vain.") Kepler was appointed to succeed him, and Tycho's instruments and hoard of observations fell into his hands. At last able to work freely, Kepler returned with a will to his studies of Mars.

Unlike Tycho, Kepler had always been a confirmed Copernican. But where Copernicus had taken the mean Sun rather than the true Sun as the center of the planetary motions, Kepler at once corrected him by making the plane of the Earth's orbit pass through the true Sun. He was rewarded for his consistency with the discovery that the orbit of Mars was inclined to the Earth's by a constant angle of 1° 50' (rather than by the variable angle Copernicus had been forced to introduce), whereupon Kepler exclaimed, "Copernicus did not know his own riches!" He next attempted to recalculate the orbit of Mars by referring it, too, to the true Sun. He began by assuming that the planet's orbit was circular but that the speed along this circle was variable. Using Tycho's observations of Mars from 1587,

1591, 1593, and 1595, he struggled to empirically locate a *punctum aequans*. After numerous trials, he almost succeeded; he produced a theory in which the discrepancy with Tycho's observations was never more than eight minutes of arc at any point around the orbit. This would have satisfied most men, but Kepler had supreme confidence in Tycho's observations, and he rejected this first theory, which he henceforth referred to as his "vicarious hypothesis."

Kepler was not greatly disappointed, since he had never cared for the concept of a *punctum aequans*. Thinking in physical terms, he could not understand why the motion of a planet should take place with respect to an empty mathematical point. Instead, he believed that the planets moved owing to a force emanating from the Sun. This was only reasonable; after all, their velocity was greatest when they were nearest the Sun and slowest when they were farthest away from it. More precisely, Kepler showed that the radius vector (the line connecting the planet to the Sun) sweeps out equal areas in equal times. This concept has become known as Kepler's second law of planetary motion, though it was actually the first he discovered.

This discovery greatly simplified his calculations, but his main goal—the shape of the Martian orbit—still eluded him. After abandoning the vicarious hypothesis, he resolved to attempt to trace the shape of the orbit without any preconception as to what that might be. His first task was to reexamine the Earth's own motion around the Sun. By considering two of Tycho's observations of Mars in which the planet had been at the same place, but regarding them from the point of view of an observer on Mars rather than Earth, Kepler was able to show that Earth's orbit was as eccentric as those of the other planets. With this information at his command, he constructed tables giving exact distances and longitudes of the Sun, and then, finally, worked out the Mars-Sun distances. Since Mars takes 687 days to orbit the Sun, it will, if measured on two dates 687 days apart, have returned to the same point in its orbit; but the Earth, which completes each revolution in 365.26 days, will be at two different positions for the measurements. If one knows the angle that Mars makes relative to the Sun at these two points, one can define the Sun-Mars distance in terms of the Sun-Earth distance.

In this way Kepler was able to calculate the distance of Mars from the Sun at various points in its orbit. Each time, he found that the distance was less than it would have been had the orbit been circular. This suggested that the orbit was an oval of some kind. In order to simplify his calculations, he began to use the more tractable ellipse as an auxiliary device, on one occasion even writing to his friend, the able observer David Fabricius, that if the orbit actually were an ellipse, the mathematical problem would already have been solved by Archimedes and Apollonius. At the moment he could take the matter no further. His efforts were heroic, and at times he came close to ruining his health. Indeed, he became so worn out that he decided to take off a whole year, 1603, and seek relaxation in researches into optics.

On returning to the "war with Mars" in early 1604, Kepler decided to use Tycho's observations to plot the planet's position in its orbit at twenty-two different points. When he did so, he found that all the points fell within the eccentric circle of the vicarious hypothesis and left a crescent, or lune, on each side between the orbit and the circle itself. He was still thinking that the shape of the orbit had to be an oval of some sort. He noticed that, relative to a circular orbit of radius 1, the distance to the lune at its greatest breadth measured 1.00429. To the casual reader this means nothing, but Kepler had agonized over the orbit of Mars for six years. In his calculations using his areal law, he had made frequent use of the so-called optical equation of Mars, which gives the angle between the Sun and the center of Mars's orbit as seen from the Earth; at this moment he happened to notice (by sheer chance, he said, but his mind had been well prepared) that for the maximum value of this angle, $5°\ 18'$, the ratio known as the secant is equal to 1.00429. This was the breakthrough he had needed. "I awoke as from a sleep," he exclaimed, "a new light broke upon me." He now grasped the relationship between the center of Mars's orbit and the distance of Mars from the Sun at the lune's widest point, and assumed that this relationship must hold true for any point in the orbit. After a few more trials, he came to his great discovery: the equation that correctly describes the orbit of Mars is that of an ellipse, with the Sun at one focus.

Kepler realized that what held true for Mars must hold true for the other planets: they too must follow elliptical paths. But he also had been fortunate; had he worked on the motions of any other planet, he would never have made this discovery. Apart from Mercury, which is difficult to observe because of its proximity to the Sun, Mars has the most eccentric orbit of the planets known during Kepler's time. Had he begun, say, with the motion of Venus, whose orbit is nearly circular, the solution would doubtless have remained beyond his grasp. Thus Kepler saw it as nothing less than providential that when he had joined Tycho in Prague, Longomontanus had been working on Mars. "In order to be able to arrive at understanding," he wrote, "it was absolutely necessary to take the motion of Mars as the basis, otherwise these secrets would have remained eternally hidden."[5]

Kepler had discovered his first two laws of planetary motion by 1605, and had also finished his great book, *New Astronomy . . . Commentaries on the Motions of Mars.*[6] To Emperor Rudolph he announced in a fittingly martial metaphor the triumph to which he had been led by his unfailing faith in Tycho's observations and his own relentless efforts:

> I bring to Your Majesty a noble prisoner whom I have captured in the difficult and wearisome war entered upon under Your auspices. . . . Hitherto, no one had more completely got the better of human inventions; in vain did astronomers prepare everything for the battle; in vain did they draw upon all their resources and put all their troops in the field. Mars, making game of their efforts, destroyed their machines and ruined their experiments; unperturbed, he took refuge in the impenetrable secrecy of his empire, and concealed his masterly progress from the pursuits of the enemy. . . .
>
> For my part, I must, above all, praise the activity and devotion of the valiant captain Tycho Brahe, who, under the auspices of Frederik and Christian, sovereigns of Denmark, and then under the auspices of Your Majesty, every night throughout twenty successive years studied almost without respite all the habits of the enemy, exposing the plans of his campaign and discovering the mysteries of his progress. The observations, which he bequeathed to me, have greatly helped to banish the vague and indefinite fear that one experiences when first confronted by an unknown enemy.[7]

Kepler hoped to win support from Rudolph to extend his investigations to the other planets. Rudolph was chronically short of funds, however, and did not have enough money to fight all his battles on Earth, let alone among the stars; even the money to publish Kepler's book on Mars was not immediately forthcoming, and its appearance was delayed until 1609.

The rest of Kepler's life was full of trials. His salary was continually in arrears, his first wife suffered from epileptic seizures and finally died, and his three children succumbed to smallpox. By 1612 Prague itself had become a battleground, and Kepler fled to Linz, in Austria. In spite of it all, Kepler continued his laborious calculations, and in 1619 announced the discovery of his third law of planetary motion—the so-called harmonic law: "The square of the period of revolution is proportional to the cube of the mean distance from the Sun."

The importance of this law is that it holds the key to the scale of the solar system. The relative distances of the planets from the Sun can be determined from their periods. If the periods are expressed in Earth years, the distances follow immediately in terms of the Earth-Sun distance (which equals 1 astronomical unit, or 1 a.u.; thus for Mars, which has a period of 1.881 Earth years and lies at a mean distance from the Sun of 1.524 a.u., $1.881^2 = 1.524^3 = 3.538$).

In 1626, Linz came under siege and Kepler was forced to flee again. Eventually he found refuge at the court of the general-in-chief of the armies of the Holy Roman Empire, Albrecht von Wallenstein, at his newly formed duchy of Sagan, in Silesia. A year later, Kepler published his long-awaited tables of planetary motion, the *Rudolphine Tables,* named for his former patron, who had died in 1612. But chronic financial worries and overwork were beginning to tell on him. Finally, in October 1630, he set out on a trip from Sagan to Regensburg, where he hoped to confer with the emperor about yet another residence, but the trip proved too much for him, and after a short illness he died on November 15, 1630.

KEPLER'S LAWS contain essential facts about the planetary motions. However, they were empirically derived from Tycho Brahe's observations. It was not until later that Isaac Newton, in his majestic *Principia* of 1687, was able to derive them from a physical theory—his principle of universal gravitation. According to Newton, every body in the universe attracts every other body with a force that is proportional to the amount of matter (mass) each contains and to the inverse square of the distance between them. In the simple two-body case, such as a planet orbiting the Sun or a satellite orbiting its primary planet, the motion of the one around the other is essentially a Keplerian ellipse; but of course matters are not so simple—every planet disturbs the motion of every other, so that when one gets down to the details the actual motions are very complicated.

Mars's motions are now well known. The elliptical path in which it moves is such that its distance from the Sun varies from 206.5 million kilometers at its closest (perihelion) to 249.1 million kilometers at its farthest (aphelion). The mean distance is 227.9 million kilometers. The planet completes each revolution in about 687 days—686.98 days to be exact.

Because of the gravitational pull of the Sun and planets on the tidal bulge in the equator of Mars, its orbit gradually changes over time; the position of its perihelion slowly rotates in space, and the shape of its ellipse is also variable—the current value of the eccentricity is 0.093 (compared with 0.017 for the Earth), but over a period of two million years it ranges between 0.00 and 0.13. I shall have more to say about the consequences of these orbital variations later.

At opposition, Mars and the Earth lie on the same side of their orbits from the Sun, and the two planets make their closest approaches to one another (because of the slight tilt of Mars's orbit relative to that of the Earth, the closest approach of the planet may actually occur as much as ten days from opposition). Since the Earth completes each orbit around the Sun in 365.26 days and Mars in 686.98 days, the Earth will overtake and pass Mars on an average of once every 779.74 days (this is known as the synodic period; the actual interval between oppositions may, however, be as little as 764 days and as much as 810 days).

If the meeting occurs when Mars is near perihelion, the distance of approach will be only 35 million miles (56 million km); if it occurs when Mars is near aphelion, the distance will be more than 61 million miles (100 million km). Since the time between oppositions is longer than the Martian year, successive oppositions are displaced around the orbit of Mars—hence the perihelic oppositions are separated from one another by an interval of fifteen or seventeen years (see appendixes 1 and 2). The last perihelic opposition was on September 28, 1988, when the minimum distance was 36,545,600 miles (58,812,900 km); the next will be on August 28,

2003, when Mars will make a closer approach to the Earth than at any time in the last several thousand years—it will come within 34,645,500 miles (55,756,600 km).

The orientation of the Martian orbit in space is such that the longitude of its perihelion currently lies at 336.06°, that is, in the direction of the constellation Aquarius. The Earth passes this point in space in late August each year, and thus perihelic oppositions always occur in August or September, when Mars is either in Aquarius itself or in nearby Capricorn. The planet then lies well to the south of the celestial equator, so that these oppositions are best observed from southerly latitudes. The reverse is true of the aphelic oppositions, which occur around the time the Earth passes the Martian aphelion (in Leo) in late February—these are best seen from the Northern Hemisphere. The difference in the size of Mars is significant; the apparent diameter of the disk ranges from 25.1″ at the perihelic oppositions to only 13.8″ at the aphelic ones. It follows that the perihelic oppositions—1877, 1892, 1909, 1924, 1939, 1956, 1971, 1988, etc.—represent the most favorable opportunities for the study of the planet, and these have generally been banner years in the history of Martian exploration.

THOUGH THE motions of Mars were worked out accurately by Kepler, nothing had yet been learned about Mars itself—a telescope was needed for that. By a strange coincidence, in 1609, the same year Kepler published his *Commentaries on the Motions of Mars*, the first telescopes were turned toward the sky. For the first time astronomers could begin to ponder not only how the planets moved but also what kind of worlds they were. A new era of Martian research had begun.

Pioneers

In 1609, after hearing reports from Holland of an invention by means of which "visible objects, though very distant from the eye of the observer, were seen distinctly as if nearby," Galileo Galilei, a professor of mathematics at the University of Padua in the Republic of Venice, was able to work out the principles of the invention for himself. He arranged two spectacle lenses in a lead tube to produce a crude telescope that magnified distant objects by a magnitude of 3×, and by August 1609 had managed to put together an instrument that magnified 8× or 10×, which he was showing off to the senators and other notables from atop the highest campaniles in Venice.

At first Galileo seems to have been interested mainly in the potential commercial applications of the new instrument. Not until November 1609, with a new telescope magnifying 20×, did he attempt a celestial application. On the evening of November 30, 1609, he turned this telescope toward the four-day-old Moon and saw that the lunar surface was rough and uneven, full of great cavities and mountains. By January 1610, he had made the first observations of the satellites of Jupiter, and also had shown that Venus could assume the crescent shape—something that was possible in the Copernican theory but not the Ptolemaic. He announced these discoveries in a lively little book, *Sidereus Nuncius* (Starry messenger), which he wrote at white heat and published in March 1610.[1] Meanwhile, he had turned his telescope for the first time to Mars, which was then near the Sun and at almost its greatest distance from the Earth.

Ever since 1597, Galileo had been a confirmed Copernican. Copernicus had shown that Mars must move around the Sun outside the orbit of the Earth. That being the case, it could not show a crescent phase as Venus does, but it could have a gibbous phase—and indeed, the maximum phase, 47°, occurs when Earth is at its greatest angular distance from the Sun as seen from Mars. The planet then appears roughly like the Moon three or four days from full. Throughout 1610 Galileo examined Mars carefully with his small telescope, hoping to make out a phase. Though the disk was barely visible and the results remained inconclusive, he wrote tentatively to Father Benedetto Castelli, one of his former pupils, on December 30, 1610: "I ought not to claim that I can see the phases of Mars; however, unless I am deceiving myself, I believe I have already seen that it is not perfectly round."[2]

By then, Galileo had made yet another discovery, which he announced in anagram form to his correspondents, who included the Jesuit fathers at the Collegio Romano in Rome and Johannes Kepler in Prague. The anagram consisted of the following letters:

s m a i s m r m i l m e p o e t a l e u m i b u n e n u g t t a u i r a s

Earlier, Kepler had congratulated Galileo on his discovery of the Jovian satellites: "I am so far from disbelieving in the discovery . . . that I long for a telescope, to anticipate you, if possible, in discovering two around Mars (as the proportion seems to require), six or eight around Saturn, and perhaps one each round Mercury and Venus."[3] He assumed that Galileo's latest discovery had to do with Mars and transposed the letters to read

Salue umbistineum geminatum Martia proles.
Hail, twin companionship, children of Mars.[4]

But Kepler misconstrued the message; in fact, Galileo's anagram concerned Saturn and was correctly rearranged as

Altissimum planetam tergeminum observavi.[5]
I have observed the most distant planet to have a triple form.

It was, in fact, the announcement of an astronomical enigma that would take half a century to solve. The triple form was actually Galileo's imperfect view of what was later shown to be the ring system of Saturn. Nevertheless, Kepler's two satellites of Mars would live on. His account may have inspired the later fancies of Jonathan Swift and Voltaire; and indeed, as we now know, the planet actually does have two satellites, though Kepler had no way of knowing this and merely made a lucky guess.

IN THE 20× instrument that Galileo used to make his first astronomical discoveries, Mars, even when closest to the Earth, would appear about the same size as a pea held at a distance of 8 feet (2.4 m). This and other early telescopes used a simple convex lens as the objective and a concave lens as the eyepiece, and had an inconveniently small field of view even at low power. Moreover, they suffered badly from spherical and chromatic aberrations. When light passes through a lens of spherical curvature, the rays near the periphery tend to focus at a point closer to the lens than those near the center—thus the image can never be brought into sharp focus. This is spherical aberration. Chromatic aberration occurs because light passing through the lens is broken into all the colors of the spectrum, and the different colors come to a focus at different points. This produces prismatic splendors around bright objects such as the Moon or Venus. Both types of aberrations are more noticeable in the light that passes through the outer parts of the lens. For this reason—and also because his lenses were by no means accurately figured—Galileo resorted to using cardboard rings in front of the object glasses so that the light would pass through only the central part of the lens, where the figure was most uniform.

Others attempted to improve the telescope and rid it of these faults. There was tremendous enthusiasm at the time, and many hoped to emulate Galileo's discoveries. As early as 1611, Kepler, in his *Dioptrice,* had proposed using a convex instead of a concave lens for the eyepiece. This is the basic idea of what became known as the astronomical telescope (as opposed to the Galilean, or Dutch, telescope). The field of view is much larger in the astronomical telescope, and although the image is inverted, this is of little importance in astronomical observations; in any case, that can be rectified by adding another lens. Unfortunately, Kepler did not actually attempt to build such a telescope, and his ideas remained little known. Apparently, the first person to actually use an astronomical telescope was a Jesuit astronomer named Christoph Scheiner, in 1617; other telescopes were built during the 1630s by Francesco Fontana, a Neapolitan lawyer and keen amateur astronomer.

With one of his telescopes, Fontana in 1636 produced a crude drawing of Mars. In this drawing the disk is shown as perfectly circular, and at the center is a dark spot which he described as looking like a "black pill." This black pill has sometimes been taken to represent one of the actual spots on the surface of the planet, but not so. Fontana later drew a similar black spot on Venus (he even used the same word, *pill,* to describe it), so there can be no doubt that the spot was the result of an optical defect in his telescope. Two years later, on August 24, 1638, Fontana made another drawing of Mars. The disk is shown as markedly gibbous in this drawing—much more so, in fact, than the planet can ever actually appear to be—and the black pill appears again, with a phase proportional to that of the disk.[6]

Although Fontana accomplished little of lasting significance, his drawings of Mars mark the beginning of what Camille Flammarion, in *La Planète Mars,* called the first period of the history of the planet.[7] The drawings made during this first period, which lasted until 1830, are rudimentary and give no real idea of the physical constitution of the planet. We may marvel at the observers' slow progress, but we must also remember that Mars is a difficult object to observe. It is a small planet—its diameter is 4,238 miles (6,780 km), only one-half that of the Earth—and it is always more than 140 times farther away from Earth than the Moon. Except during the periods when it is very near the Earth, its disk is always small, and details are not easy to see.

Though Fontana cannot have made out any of the actual spots on Mars, others using better telescopes succeeded in doing so. A tantalizing glimpse was obtained by a Neapolitan Jesuit named Father Bartoli, who on December 24, 1644, described two patches on the lower part of the disk. More observations of patches were reported in 1651, 1653, and especially in July and August 1655, when Mars was near a perihelic opposition, by Giambattista Riccioli and Francesco Grimaldi, Jesuits of the Collegio Romano. But the true credit must go to Christiaan Huygens, the great Dutch astronomer (fig. 1).

FIGURE 1. Christiaan Huygens.
(Courtesy Yerkes Observatory Photographs)

BEGINNING IN 1655, when he was twenty-six, Huygens began experimenting with new ways of figuring lenses for microscopes and telescopes. In the course of these experiments, he devised the first compound eyepiece, still known as the Huygenian. By March 1655 he had constructed a good 2-inch (5.1-cm) telescope of 10.5-foot (3.2-m) focal length and 50× magnification, with which he discovered Saturn's largest moon, now known as Titan. Soon afterward he was able to solve the enigma introduced by Galileo by showing that Saturn was surrounded by a ring.[8] He also observed Mars, but though it had just passed a perihelic opposition in July 1655, he apparently did not get around to it until long after opposition—he could make out nothing on the disk except that it appeared to be "crossed by a sombre band."

Busy with other projects, including perfecting a pendulum clock and writing his book on the ring of Saturn, Huygens made no other observations of Mars until 1659. On November 28, at 7:00 P.M., he turned his telescope toward the planet, which was then near opposition and showing a disk 17.3″ across. There were patches on the disk, and Huygens made a sketch showing a V-shaped marking, which, to anyone with the least knowledge of Mars, is immediately recognizable as Syrtis Major (for a long time known rather more descriptively as the Hourglass Sea). Huygens was able to detect a slight shift in the marking's position during the time he kept it under observation, and when he turned his telescope toward the planet again, on December 1, he found that it had returned to very nearly the same place on the disk. Thus he noted in his journal: "The rotation of Mars, like that of the Earth, seems to have a period of 24 hours."[9]

HUYGENS'S MAIN rival for telescopic glory at the time was Giovanni Domenico Cassini (fig. 2). Cassini was born in 1625 at Perinaldo, near Nice, though, according to Flammarion, he was "by temperament much more Italian than French."[10] The French writer Bernard le Bovier de Fontenelle would later pay him the ultimate compliment by linking him with Galileo: "These two great men," he wrote, "made so many discoveries in the sky that they resemble Tiresias, who lost his sight for having seen some secrets of the gods"—an allusion to the fact that both Galileo and Cassini were blind in old age.

In 1648, Cassini was invited by Marquis Cornelio Malvasia, a wealthy amateur astronomer, to come to work at his private observatory at Panzano, near Bologna. This gave him access to telescopes with which he could begin to do useful research, and at the same time he completed his education under the Jesuits Riccioli and Grimaldi. Two years later, he succeeded Cavalieri in the chair of astronomy at Bologna University and became acquainted with the skillful Roman instrument maker Giuseppe Campani.

In 1664, Cassini turned a Campani telescope of 17-foot (5.2-m) focal length toward the planets, with remarkable results. On Jupiter he made out not only the dark belts but also various temporary spots, from which he worked out the rotation period at just under ten hours. He also paid attention to the Jovian satellites' eclipses and shadow transits, from which he drew up accurate tables of their motions. In 1666, he recorded spots on Venus and concluded—rather

FIGURE 2. Giovanni Domenico Cassini.
(Courtesy Yerkes Observatory Photographs)

ambiguously—that it "completed its movement of either revolution or libration in less than one day, so that in twenty-three days, approximately, it shows the same aspect." [11]

During the same period, Cassini detected patches on the surface of Mars, despite the fact that the planet was not very favorably placed for observation—its opposition, on March 19, 1666, was an aphelic one. His drawings are primitive, and the spots, of which the most conspicuous are represented rather in the form of a dumbbell, are not easily related to those actually known to exist on the planet. Like Huygens, he noted their slow drift across the disk, and found that after a period of thirty-six or thirty-seven days they returned to the same positions at the same hour of the night, from which he was able to work out the rotation period at 24 hours, 40 minutes. Since Huygens's earlier results had never been published, Cassini's determination was completely independent, and it is very close to the actual value. This proves, if any proof were needed, that the markings he was following cannot have been illusory.[12]

The same opposition was observed by Robert Hooke, of the Royal Society of London, on the eve of the Great Fire that destroyed much of the city that year. In the 36-foot (11-m) telescope he was using, Hooke found that Mars was very nearly as large as the Moon viewed with the naked eye. Nevertheless, he wrote, he had a difficult time because of the unsteadiness of the air:

> Such had been the ill disposition of the Air for several nights, that from more than 20 observations of it, which I had made since its being Retrograde, I could find nothing of satisfaction, though I often imagin'd, I saw Spots, yet the Inflective veins of the Air (if I may so call those parts, which, being interspers'd up and down in it, have a greater or less Refractive power, than the Air next adjoyning, with which they are mixt) did make it confus'd and glaring, that I could not conclude upon any thing.[13]

He persisted and eventually enjoyed a few nights of transparent and steady air, so that the disk of Mars became "so very well defined, and round, and distinct." Under these conditions he was able to make out some of the spots clearly, and his drawings, though primitive by any standard, do in a few cases show Syrtis Major and other Martian features in recognizable form.

Obviously, the telescopes of the late 1600s gave woefully inadequate definition. Moreover, another factor had entered into the discussion, which nowadays we refer to as the "seeing," the atmospheric conditions under which observations are made. Hooke had clearly described its role in planetary observation, and Huygens was also aware of it. He noted that the stars twinkled and that the edges of the Moon and planets trembled in the telescope, even when to all appearances the atmosphere appeared calm and serene. So frequent were the bad nights that Huygens warned against too hastily blaming the telescope for indifferent results.[14]

WOOED BY Louis XIV, who wanted to add to the renown of the Académie des Sciences, Huygens left Holland for Paris in 1666 and took up quarters in the Bibliothèque de Roi. He would remain in France as the most prestigious member of the Académie for the next fifteen years. Three years later, Cassini was also drawn to France by the Sun King in order to lead the newly founded Paris Observatory.

On first arriving in Paris, Cassini found that detailed plans for the observatory had already been drawn up by Claude Perrault, the architect responsible for designing the new façade at the Louvre. Cassini, who had an astronomer's eye rather than an architect's, objected strongly to the plans. Eventually a meeting was arranged between Cassini and Perrault, with the king and his finance minister, Jean-Baptiste Colbert, also on hand. Cassini's great-grandson Jean Dominique Cassini IV later described the meeting:

> Perrault eloquently defended his plan and architectural style with beautiful sentences. My great-grandfather spoke French very poorly, and in defending the cause of astronomy he shocked the ears of the King, Colbert, and Perrault to such a point that Perrault in the zeal of his defense said to the King: "Sire, this windbag doesn't know what he is talking about." My great-grandfather kept silent and did well. The King agreed with Perrault and did badly. The result is that the observatory has no common sense.[15]

Undaunted, Cassini set up the 17-foot (5.2-m) Campani telescope he had brought with him from Italy in the courtyard outside the observatory and set to work—with brilliant results. In 1671, with this telescope, he discovered a satellite of Saturn, Iapetus. A year later, with a 34-foot (10.4-m) Campani telescope, he added another, Rhea. Even these telescopes must have been difficult to handle; Cassini mounted the Campani lenses in light wooden tubes, which he suspended from a high mast on the observatory terrace. Later, he began using even longer telescopes, and in 1684 he discovered two more satellites of Saturn, Dione and Tethys, with Campani telescopes of 100- and 136-foot (30.4- and 41.5-m) focal lengths mounted atop an old wooden water tower which he had had transported to the observatory. The tower was equipped with a stairway and also had a balcony around the top to prevent his assistants from falling off on dark nights!

The two greatest observational astronomers of the age sometimes observed together, and both made important observations of Mars at the perihelic opposition of September 1672. Huygens made another drawing of the planet which shows Syrtis Major unmistakably, and also drew the first clear representation of the brilliant south polar cap. (It is often said that Cassini deserves credit for the discovery of the polar caps because one of his drawings from 1666 shows bright patches at the poles; however, I am not convinced, since the same drawing also shows bright patches of the same sort at both limbs.) Cassini concerned himself less with making physical observations of the planet than with measuring its position relative to the stars. By comparing his results with those obtained by another French astronomer, Jean Richer, who had traveled to Cayenne, French Guiana, Cassini worked out the parallax of Mars, which in 1672 was two and a half times greater than that of the Sun. This gave the distance to Mars, and from Kepler's harmonic law (which, as noted earlier, relates the distances of the planets to their periods of revolution) he was able to work out the value of the astronomical unit, the distance from the Earth to the Sun. In fact, there were large errors in the measurements, and Cassini was lucky that the distance he calculated, 87 million miles (140 million km), was as close to the actual value (92,955,800 miles, or 149,597,870 km) as it was.[16]

In the 1680s, Huygens returned to Holland to escape the persecution of Protestants that had arisen in an increasingly militant Catholic France. (Cassini, who remained a good Catholic, was never in danger; indeed, to the end of his life—he died in 1712—he remained a supporter of the Tychonic system rather than the Copernican, which had been condemned by the church.) In Holland, Huygens settled on a country estate at Hofwjick, near The Hague, and continued his efforts to improve the telescope.

Chromatic aberration had been the plague of observers from Galileo's time onward. Instrument makers naturally wanted to produce telescopes with larger apertures, but when they tried to do so, they found that the chromatic aberration became even worse. Eventually, it was found that by making the curvature of the lens shallower and its focal length longer, the effect of chromatic aberration could be reduced. Thus telescopes grew longer and longer. Huygens produced some of the first long telescopes, and his discoveries inspired the even longer telescopes of Johannes Hevelius, a brewer and city councillor of Danzig (now Gdansk), which reached lengths of 60, 70, and even 150 feet (18.2, 21.3, and 45.7 m); they were destroyed by the great fire at Danzig in 1679. Such telescopes were extremely unwieldy and difficult to use, and Huygens himself decided to turn his attention to tubeless, or "aerial," telescopes, in which the object glass was fixed to the top of a tall mast and the observer sighted along guy wires, which could be used to point the object glass in any desired direction. Bright objects such as planets could be found either by receiving them on a white pasteboard ring fixed around the eyepiece, which the observer held by hand, or, more conveniently, by receiving them on an oiled and translucent paper screen (fainter objects obviously posed much greater difficulties). The object glass was illuminated by a lantern, and the observer searched for the lantern's reflection in order to bring the lenses into alignment. In 1686, Huygens produced objectives with diameters of 7.5, 8, and 8.5 inches (19, 20, and 22 cm); their focal lengths were 123, 170, and 210 feet, respectively (37.5, 51.8, and 64 m). He used them to observe the perihelic opposition of Mars in 1686, though his sketches show no more detail than that recorded in his earlier sketches made with more modest instruments. His very last sketch of the planet was made on February 4, 1694, with Mars at an aphelic opposition. Huygens died in 1695.

WITH HUYGENS'S passing, the great century of discovery came to a close. Astronomy had made tremendous strides, though relatively little had been learned about Mars. It was known to have various dark and light patches, and there were hints that these might change over time. As early as 1666, Cassini had suggested that, seen from a great distance, our globe would resemble the other planets. Following up a remark by Galileo, he had predicted that the seas would appear dark because they absorb sunlight, and the continents would appear bright. However, he stopped short of applying this explanation to the perceived differences in the Martian globe. In 1686, Bernard le Bovier de Fontenelle first published his charming *Entretiens sur la pluralité des mondes* (Conversations on the plurality of worlds), in which he speculated about the conditions of life and the nature of the inhabitants of the Moon, Mercury, and Venus. He gave cursory attention to Mars, however, saying only that "Mars has nothing curious that I know of; its days are not quite an hour longer than ours, and its years the value of two of ours. It's smaller than the Earth, it sees the Sun a little less large and bright than we see it; in sum, Mars isn't worth the trouble of stopping there."[17] A far cry from the rich speculations that would gather around the planet in succeeding centuries! As yet, Mars, seen through the primitive telescopes of that era, had proved too unrewarding an object to pique much interest.

In the last years of his life, Huygens attempted to formulate his own ideas about extraterrestrial life, and "as they came into his head . . . clapt them down into common places."[18] The result was his *Kosmotheoros,* which was finished by January 1695, though its author's death six months later delayed its publication until 1698. Huygens declared that the planets must have vegetation and animals, because without such life "we should sink them below the Earth in Beauty and Dignity; a thing that no Reason will permit."[19] Though Mars, because of its greater distance from the Sun, would be much cooler than the Earth, Huygens felt that the inhabitants would be adapted to their conditions. Its rotation, he declared, was established without question from the movements of its spots, and proved to be similar to that of the Earth, while its axis seemed to be only slightly inclined to the plane of its orbit, so that there would not be much difference in the seasons for its inhabitants. Beyond this, there was little more to be said about the planet.

The situation, alas, was unlikely to change very rapidly. The aerial telescopes used by Huygens in the last years of the seventeenth century admitted little development beyond what they had achieved in their inventor's hands, and they were cumbersome and difficult to use. Huygens had bequeathed his aerial telescopes to the Royal Society of London, but they were scarcely ever used. The 123-foot (37.5-m) telescope was occasionally dusted off, but the results were not encouraging. "Those here that first tried to make use of this Glass," one of the members wrote in a note to the *Philosophical Transactions of the Royal Society* in 1718,

> finding for want of Practice, some difficulties in the Management thereof, were the occasion of its being laid aside for some time. Afterwards it was designed for making perpendicular Observations of the fixt Stars passing by our Zenith, to try if the

Parallax of the *Earths* annual Orb might not be made sensible in so great a Radius, according to what Dr *Hook* had long since proposed: but in this we miscarried also, for want of a place of sufficient height and firmness, whereon to fix the Object Glass, so that it lay by neglected for many Years.[20]

Clearly, aerial telescopes did not encourage protracted labor at the eyepiece, and it is little wonder that results were so meager over the next three quarters of a century—a period that deserves to be regarded as the "long night" of Martian studies.

"A Situation Similar to Ours"

When Giovanni Domenico Cassini became blind in 1710 (he died two years later), his son Jacques took charge of the Paris Observatory. Indeed, the Cassinis were destined to become one of the most remarkable astronomical families in history. Jacques (Cassini II) was succeeded by his son, César-François Cassini, who is also known as Cassini de Thury (Cassini III), who in turn was succeeded by his son, Jacques Dominique Cassini (Cassini IV). In politics, the family remained strongly royalist in its sympathies, and Cassini IV was finally forced to resign in 1793, during the height of the French Revolution.

Among the other worthies in this illustrious family's tree was the elder Cassini's nephew, Giacomo Filippo Maraldi (born in 1665), who became an assistant at the Paris Observatory. Although overshadowed by the fame of his brilliant uncle, Maraldi was a competent astronomer in his own right; his work included the compilation of a star catalog and the calculation of cometary orbits. But he will always be best remembered for his work on Mars, which he pursued almost singlehandedly during a period when studies of the planet were generally neglected.

Maraldi made a careful series of observations at every opposition beginning in 1672; his best results were obtained at the perihelic oppositions of 1704 and 1719.[1] He described the markings on Mars as not normally well defined even in large telescopes—most of his observations were made with the 34-foot (10.4-m) Campani telescope of the Paris Observatory—and suspected that they were variable not only from opposition to opposition but even from month to month. His sketches show a dusky band near the middle of the disk, which reminded him of one of the cloud belts of Jupiter; it was interrupted in places and occupied only somewhat more than one hemisphere of Mars. At one point this band was joined by another at an oblique angle, and taking this as a reference point, Maraldi was able to determine the rotation of the planet. This point seemed to return to the same location on the disk after thirty-seven days, during which the planet had rotated thirty-six times. He therefore worked out the rotation period at 24 hours, 40 minutes, which agreed exactly with his uncle's result. In addition to the bands just described, he made out a large triangular patch on the Martian surface. The features described by Maraldi are readily identified on modern maps: the band consists of the darkish swath from Mare Sirenum in the east to Mare Tyrrhenum in the west, while the dark triangular patch can be nothing other than Syrtis Major, the "Hourglass Sea." Nevertheless, Maraldi himself was not convinced of their permanence; he thought they were merely clouds.

Though the south polar spot had been sketched by Huygens in 1672, it was Maraldi who began the first thorough study of the Martian poles. He found both poles to be marked by whitish spots, though because the southern hemisphere of Mars was tilted toward the Earth at the perihelic oppositions, the spot near the south pole was markedly easier to observe. When the polar spot was very small, Maraldi noted that it underwent a small revolution as Mars rotated, which meant that its position was slightly eccentric to the pole. Moreover, there were changes in the spot's extent: in August and September 1719 it disappeared entirely, though later it returned. This behavior suggested that the material of which the spot was made underwent physical changes of some kind, although Maraldi himself declined to speculate on what they might be.

AFTER MARALDI'S observations of 1719, little work was done on Mars for the next several decades, and of that little, none is deserving of special notice. We pass by, therefore, the wasted perihelic oppositions of 1734, 1751, and 1766, and go directly to the contributions of William Herschel, one of the greatest astronomers who ever lived. Though his most important work concerned the sidereal universe, in his early career Herschel made numerous observations of the Moon and planets.

Friedrich Wilhelm Herschel (who later naturalized his name to William) was born at Hanover, in Germany, in 1738. His family was musical, and at first young Herschel attempted to follow his father into a career as a bandsman in the Hanoverian Guard. In the spring of 1757, however, after the Guard's disastrous campaign against the French during the Seven Years' War (Herschel himself came under fire at the Battle of Hastenbeck), he decided to leave the Guard and seek his fortune in England. England and Hanover had enjoyed close ties ever since 1714, when a Hanoverian prince, George Louis, ascended the throne of England as George I. Many Germans followed the Georges to the court of London, and Herschel was drawn like a magnet. He struggled at first, but by 1766 had established himself as a musician in the dazzling resort city of Bath—he was the organist for the Octagon Chapel, and also composed music and gave private lessons. Though for the next ten years he made his living as a busy musician, his interests turned increasingly to astronomy, and before long he began to cast about for a suitable telescope.

The refractor, as we have seen, had reached an apparent dead end with the aerial telescopes of Huygens and Cassini. Isaac Newton, the architect of the theory of gravitation, was among those who concluded that the problems were insurmountable, and he proposed that a curved mirror be used in place of the lens to collect the light. In the Newtonian reflector—so-called to distinguish it from the many other variations on the basic idea—the mirror reflects the light back up the tube to a small, flat mirror, which is set at a 45° angle so as to redirect the beam through a hole in the side of the tube, where the image can be magnified by an eyepiece. Provided the curve given to the main mirror is a parabola, all the light is brought to a single focus; and since the light never has to pass through a lens until it reaches the eyepiece, chromatic aberration can be avoided. Newton himself actually produced a reflector with a 1-inch (2.5-cm) mirror made of bell metal, which he presented to the Royal Society of London in 1672. Unfortunately, the metals used in the mirrors of the day were difficult to figure and bring to a good polish, and it was not until 1722 that John Hadley was able to produce a reflector capable of equaling the performance of Huygens's aerial telescopes.

In 1773, when Herschel began to take a serious interest in astronomy, he first directed his attention to refractors, but he found the long tubes almost impossible to manage. He next rented a small reflector and, finding it satisfactory, attempted to purchase such an instrument. Unfortunately, those available were beyond his means, and he decided to experiment with mirror making. By this time his sister Caroline and brother Alexander had come over from Hanover and joined him in Bath, and every room in the house in which they were then living soon took on the appearance of a workshop. After a number of failures at mirror making, Herschel finally, in 1774, succeeded in making a small reflector with which he recorded his first observation, of the Orion nebula. He went on to produce larger and larger instruments that were far superior to any others of his day. In 1777, he moved to 19 New King Street near the center of Bath, bringing with him working reflectors of 7-foot (2.1-m) and 9-foot (2.7-m) focal lengths. From the long, south-facing garden in back of the house he made a few observations of Mars at its opposition that year—in particular, he recorded the "two remarkable bright spots on Mars" (the polar caps).

HERSCHEL RETURNED to Mars in 1779, and again in 1781. On the most fateful night of his career—March 13, 1781—he observed Mars with a recently completed 20-foot (6.1-m) reflector and recorded in his notebook that there was a "very lucid spot on the southern limb . . . of a considerable extent." It was on that same night, between 10:00 and 11:00 P.M., he had discovered with his 7-foot telescope a tiny disk among the stars. At first he thought it was a comet, but the disk later proved to be something much more consequential: it was nothing less than a new planet, the first discovered in modern times.

This discovery changed Herschel's life. He was granted a pension by George III that allowed him henceforth to spend all of his time on astronomy. In gratitude, Herschel proposed to call the planet Georgium Sidus, the Star of George, but the name never stuck. On the Continent (and eventually, after Herschel's death, in England also) it was superseded by the name Uranus, which had been proposed by the German astronomer Johann Elert Bode.

In the months after his great discovery, Herschel's attention was not entirely absorbed in observing the new planet. He did return to Mars from time to time as it came to a perihelic opposition on July 27, 1781, and from his observations worked out a new rotation period: 24 hours, 39 minutes, 21.67 seconds, which is exactly 2 minutes too long. He was also able to confirm that the north polar spot, which he observed carefully during the months of June and July, was eccentric to the pole—the same result that Maraldi had earlier established for the south polar spot.[2]

Before the next opposition of Mars, in 1783, William and Caroline moved from Bath to Datchet, near Windsor. There William rented a dilapidated old house not far from the Thames. It was large and commanded an excellent view of the sky, and it was reasonably close to Windsor Castle in case the royal family should want to look through his telescopes. There William worked incessantly, as his sister later recalled:

> The assiduity with which the measurements on the diameter of the Georgium Sidus, and observations of other planets, double stars, etc., etc., were made, was incredible, as may be seen by the various papers that were given to the Royal Society in 1783, which papers were written in the day-time, or when cloudy nights interfered. Besides this, the twelve-inch speculum was perfected before spring, and many hours were spent at the turning bench, as not a night clear enough for observing ever passed but that some improvements were planned for perfecting the mounting and motions of the various instruments then in use, or some trials were made of new constructed eye-pieces, which were mostly executed by my brother's own hands.[3]

In September and October 1783, Herschel observed Mars extensively—the planet came to a very good almost-perihelic opposition on October 1—and noted that the south polar cap was very small. On October 1, he jotted in his notebook: "I am inclined to think that the white spot has some little revolution. . . . It is rather probable that the real pole, though within the spot, may lie near the circumference of it, or one-third of its diameter from one of the sides. A few days more will show it, as I shall now fix my particular attention on it."[4] Later that night he established that the spot had "a little motion, for it is now come farther onto the disk."[5] Over the next several nights he carefully measured its position from hour to hour. Later, he used these observations to calculate that the south polar cap was 8.8° from the south pole. This was not unexpected; on Earth, the pole of greatest cold does not correspond with the geographical pole. He was also able to work out the inclination of the planet's axis to the plane of its orbit—28° 42′—and fixed the equinoctial point on the Martian ecliptic (the vernal equinox of Mars) at 19° 28′, in Sagittarius. As the planet's axial inclination was very nearly the same as that of Earth, Herschel realized that the Martian seasons must be analogous to ours, though nearly twice as long. From this, he thought he could "account, in a manner which I think highly probable, for the remarkable appearances about its polar regions":

> The analogy between Mars and the earth is, perhaps, by far the greatest in the whole solar system. The diurnal motion is nearly the same; the obliquity of their respective ecliptics, on which the seasons depend, not very different; of all the superior planets the distance of Mars from the sun is by far the nearest alike to that of the earth: nor will the length of the martial year appear very different from that which we enjoy, when compared to the surprising duration of the years of Jupiter, Saturn, and the Georgium Sidus. If, then, we find that the globe we inhabit has its polar regions frozen and covered with mountains of ice and snow, that only partly melt when alternately exposed to the sun, I may well be permitted to surmise that the same causes may probably have the same effect on the globe of Mars; that the bright polar spots are owing to the vivid reflection of light from frozen regions; and that the reduction of those spots is to be ascribed to their being exposed to the sun.[6]

Herschel calculated the diameter of Mars to be 0.55 times that of the Earth, and he found its figure to be just as flattened as that of Jupiter—the ratio of the equatorial to the polar diameter he put at 16/15.

The dark markings on the planet came only incidentally within the course of Herschel's survey, and he made little more than crude sketches of them. Even so, they are far more clearly delineated in his drawings than they are in those of his predecessors. With Herschel's work, Mars studies left the Maraldi era behind and entered a new epoch. The drawings of 1783 are of special interest. One easily recognizes many Martian features: the Hourglass Sea is clearly shown, as are the features that later became known as Sinus Sabaeus and Sinus Meridiani. This alone would be sufficient to show that the Martian surface markings are generally fixed and permanent. However, Herschel's drawings also provide unmistakable evidence of changes. For example, he showed a prominent feature that is no longer present—a triangular patch, somewhat similar to Syrtis Major, which curves down from Mare Cimmerium to end almost in the form of a hook.

In addition to these studies of the surface, Herschel made a few observations with a bearing on the question of whether Mars had an atmosphere. Cassini, in 1672, had observed a fifth-magnitude star (Phi Aquarii) disappear a full six minutes from the disk of Mars, which led him to conclude that the planet must have a very dense atmosphere. Herschel suspected that rather than being hidden by the planet's atmosphere, the star had merely disappeared into the glare around Mars, and eventually he was able to put the question to a test. Using his 20-foot (6.1-m) reflector, which had a mirror 18.7 inches (47 cm) in diameter, he followed two faint stars as they approached Mars without noting the least diminution of their light. This proved that the Martian atmosphere could not be as appreciable as Cassini had supposed. Nevertheless, Herschel did have evidence that an atmosphere existed. "Besides the permanent spots on its surface," he wrote, "I have often noticed occasional changes of partial bright belts . . . and also once a darkish one, in a pretty high latitude. . . . And these alterations we can hardly ascribe to any other cause than the variable disposition of clouds and vapours floating in the atmosphere of that planet." Thus he concluded that the inhabitants of Mars "probably enjoy a situation in many respects similar to ours."[7]

AFTER THE 1783 opposition of Mars, Herschel began to wrestle with the momentous questions of the sidereal universe and the construction of the heavens, and henceforth paid scant attention to Mars. Nevertheless, his success inspired others—and none more than Johann Hieronymus Schroeter, who became one of the most enthusiastic amateur astronomers of all time and the next important observer of Mars.

Although interested in astronomy from an early age, Schroeter followed the family profession of law. He graduated from the University of Göttingen, which had been founded by George II, and in

1777 was appointed secretary of the Royal Chamber (of George III) in Hanover. Music was another of his many interests, and this brought him into contact with two of William Herschel's younger brothers, Johann Alexander and Dietrich, through whom his interest in astronomy was reawakened. Dietrich helped him to obtain a small telescope in 1779, and Schroeter used it to make a few observations of the Moon and Venus. However, the decisive event of his life took place in 1781, with William Herschel's discovery of Uranus. In a spirit of emulation, Schroeter resolved to dedicate himself henceforth to astronomy and resigned his position in Hanover in order to take on the less demanding post of chief magistrate of Lilienthal, a village on the moor near Bremen. He took up residence in the Amthaus there in 1782. Meanwhile, he had succeeded in obtaining two mirrors, 4.75 and 6.5 inches (12 and 16.5 cm) in diameter, that had been made by William Herschel himself. The larger of the two Schroeter assembled into a 7-foot (2.1-m) reflector that was in every respect identical with the one Herschel had used to discover Uranus; when Schroeter began to use it, in 1786, it was the largest telescope in Germany.

Unlike Herschel, who was mainly concerned with stellar and nebular astronomy, Schroeter's lifelong interest centered almost entirely on the Moon and the planets. He was a compulsive acquirer of telescopes—by 1793, he had erected several in the garden of the Amthaus, of which the largest had a 19.25-inch (49-cm) mirror and a focal length of 27 feet (8.2 m). He was also a tireless observer, and he published, at his own expense, a succession of large tomes describing his observations. One concerned with his lunar work, the *Selenotopographische Fragmente,* appeared in 1791, followed by a second volume in 1802. There were similar volumes devoted to each of the planets.[8]

As might be expected, Mars came in for its share of attention. Schroeter's first observations of it were made with the smaller of the two Herschel telescopes in November 1785. He noticed only a "few grey, misty, poorly bounded patches."[9] As he followed these patches from night to night, he thought they often appeared to be similar, but he could never quite convince himself that they were identical. At the next opposition, in 1787, he was able to use the larger of his Herschel telescopes on Mars for the first time, and his suspicions about the instability of the Martian surface hardened into an *idée fixe*. "The spots and streaks on the globe of Mars are always changing," he wrote, "even from hour to hour. But that they are the same regions is shown by the fact that the same shapes and positions develop and pass away again, as one would expect of the variable atmospheric appearances occurring above a solid surface."[10] Very strange, since the drawing he made on this particular night shows Syrtis Major in clearly recognizable form! The region is the same one Huygens had figured so well in 1659, so there ought to have been no question whatever of the permanence of the markings in this of all regions—Schroeter's failure to recognize it only testifies to the difficulty of perceiving an unfamiliar object correctly and the strong influence of fixed ideas. Indeed, some of his later drawings actually show the Martian markings in the form of dark belts similar to those of Jupiter. The delusion that the patches on Mars were mere cloud forms pervaded all of Schroeter's work, and though he made careful observations at all the later oppositions—especially the excellent perihelic opposition of August 30, 1798—he never doubted that what he was seeing was a mere floating shell of clouds.[11]

Schroeter also carried out careful observations of the polar caps, which he considered to be the result of a "dazzling atmospheric precipitation."[12] His measures of the diameter of the planet were very close to those of Herschel, though he determined that the equatorial and polar diameters agreed to within 1/81 (in this he was more nearly correct than Herschel; the polar and equatorial diameters are now known to agree to within 1/500). He confirmed Herschel's results as to the obliquity of the axis and the nature of the Martian seasons, and concluded that of all the planets, Mars was the most similar to Earth.[13]

Schroeter's later days were sad ones. He lived in the turbulent era of the Napoleonic Wars, and the peaceful "Vale of Lilies" was engulfed in 1806 when it came under the control of the French. Henceforth Schroeter was cut off from the financial support he had enjoyed from George III, and the French did not pay him for his work as magistrate, which included collecting taxes for them. He was soon so straitened that he found it difficult to keep up the observatory, and his situation became even worse in 1810 when he was dismissed from his position. By Napoleon's decree, Lilienthal became part of the Department de le Bouche de Weser, which had Bremen as its capital. Nevertheless, Schroeter carried on as best he could, and he was able to observe the Great Comet of 1811.

The worst was yet to come. In April 1813, as the French were reeling back from their disastrous winter campaign in Russia, a skirmish took place near Lilienthal between a French detachment and a small band of Russian Cossacks. A French officer was wounded and reported that his detachment had been fired upon by the local peasantry. Without further notice, the French general, Vandamme, gave orders to set fire to Lilienthal. A strong wind fanned the conflagration, which destroyed the government buildings where Schroeter kept many of his books and manuscripts. Schroeter himself was "obliged to fly with my family, in our night dresses, to my farm at Adolphsdorf." His observatory escaped the inferno but was broken into and plundered by French troops several days later. They "with a fury the most unprovoked and irrational destroyed or carried off the most valuable clocks, telescopes, and other astronomical instruments."[14]

Soon afterward the French were expelled from Germany, and Schroeter, reinstated as chief magistrate, attempted with all the strength left in him to rebuild Lilienthal. But it was too late for him to try to rebuild his observatory. He kept despair at bay by writing up his observations of the Great Comet of 1811, and then turned to his observations of Mars, which had never been published. Miraculously, most of those records had escaped the fire, though a few of the drawings were damaged and had to be redrafted. Schroeter's engraver, Tischbein of Bremen, began to make copper plates of the drawings, but Schroeter's eyesight was failing, and the project was still unfinished when he died in August 1816. For some reason his heirs declined to carry it through to completion.[15]

Thereafter, Schroeter's manuscripts and drawings of Mars remained unknown until 1873, when François Terby, an enthusiastic student of the red planet who had a private observatory at Louvain, Belgium, and was diligently collecting drawings of Mars for his monograph *Aréographie* (1875),[16] succeeded in tracking them down among the effects of one of Schroeter's nephews. Terby later deposited the material in the library of the University of Leyden, and Henricus Gerardus van de Sande Bakhuyzen, director of the Leyden Observatory, finally edited Schroeter's work on Mars and published it in 1881. Although the work appeared too late to have any real influence, Schroeter's drawings nevertheless remain valuable—indeed, as Bakhuyzen remarked, their worth is actually increased "because of Schroeter's erroneous view that the spots on Mars were mere cloud-forms, which sometimes changed very swiftly. For when Schroeter observed these spots, he was not predisposed to see the same details, but his different representations of them are as fully free of prejudice and independent of one another as possible."[17]

It remains surprising that Schroeter failed to recognize that the features he saw on the surface were the same. Many of his drawings show Syrtis Major, and other major features are also easily recognized; for instance, he made a vivid record of the round feature that later became known as Solis Lacus. The most interesting drawings are those that record the now-vanished curved hook, which, as mentioned earlier, Herschel first depicted in 1783, a marking Bakhuyzen referred to as "Spitze B" and Joseph Ashbrook called the "Arrowhead" (fig. 3). Located at about longitude 240° w, it extended from Mare Cimmerium into the region known on later maps as Aethiopis and was one of the most visible features on Mars during the last two decades of the eighteenth century, rivaling, and sometimes even mistaken for, Syrtis Major itself. Subsequently the hook disappeared, in what Ashbrook called "the most striking change yet recorded on the surface of the red planet."[18] The fact that its history is known at all is largely owing to the astronomer of Lilienthal and his candid records.

Before closing the chapter on Schroeter, there is one final fact to consider. We now know that from time to time much of the surface of Mars is obscured by dust storms, of which I will have much to say later. With his keen eye, large instruments, and long-sustained observations, Schroeter would have been in an ideal position to record dust storms had any occurred during the period of his watch. I have gone to the trouble of calculating the longitude of the central meridian for each of Schroeter's 231 drawings of Mars, and have concluded that there is no clear evidence of dust storm activity—a negative result which is itself of value, since it seems to indicate that there were no planet-encircling or global storms during the period covered by Schroeter's observations.

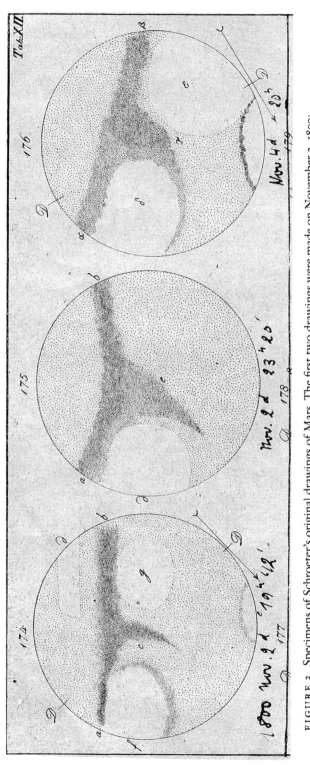

FIGURE 3. Specimens of Schroeter's original drawings of Mars. The first two drawings were made on November 2, 1800; the one on the left shows the "Arrowhead" near the central meridian; in the second drawing, made four hours later, the "Arrowhead" has rotated off the disk and Syrtis Major is on the central meridian. The drawing on the right was made on November 4, 1800, and again shows the "Arrowhead" on the central meridian. (Courtesy Dieter Gerdes)

WITH THE work of Herschel and Schroeter, the first phase of the physical study of Mars was under way. Its rotation period, axial inclination, seasons, polar caps, and atmosphere were now reasonably well known, and a tentative start had been made on the Martian geography. Whether the markings were permanent, as Herschel believed, or cloud formations, as supposed by Maraldi and Schroeter, remained the first order of business to be settled by the observers of the new century. But regardless of the outcome of those further investigations, the planet had acquired the distinction of being by all odds the most Earthlike, and thus had gained immeasurably in interest over the days when Fontenelle was able to dismiss it in a few lines.

Areographers

Mars is by no means an easy object to study. It is a small world, only slightly more than half the diameter of the Earth, and even at its nearest it never approaches closer than 140 times the distance of the Moon. Moreover, the features on its surface are of low contrast, and thus are difficult to delineate accurately. To study Mars properly, perfect instruments and a steady atmosphere on Earth are required.

The large reflectors used by Herschel and Schroeter showed considerable detail and allowed the first reasonably thorough study of the planet's polar caps, axial inclination, and seasons. These instruments were far from ideal, however; the mirrors, which were made of speculum metal—an alloy of copper and tin—were difficult to figure accurately, and they tarnished easily. (It was only in the late nineteenth century that the more satisfactory method of silver coating glass mirrors was introduced.)

Meanwhile, the refractor had begun to make a comeback. In 1729, an Englishman named Chester Moor Hall showed that Newton had been premature in considering the problem of the refractor insoluble.[1] Hall showed that when a concave lens of flint glass was combined with a convex lens of crown glass, the chromatic aberration produced by one lens could be made to nearly cancel that of the other for a certain band of wavelengths; say, the yellow and green spectral region to which the human eye is most sensitive. He had discovered the secret of the achromatic lens, but he himself did not pursue it further, and the importance of his discovery was not fully realized by other opticians. It was not until after 1750 that another Englishman, John Dollond, went into business with his son Peter to make achromatic lenses on a commercial basis.

A good refractor with a 2.5-inch (64-mm) objective lens could now have a length of 20 inches (51 cm) instead of 20 feet (6.1 m), as in the days of the aerial telescopes. Obviously, this was a tremendous improvement; but even so, the early instruments of the type were far from perfect. Their components were not correctly matched, and bright objects seen through them continued to have a troublesome amount of unfocused light (secondary spectrum), so that they were surrounded by green and wine-colored fringes. Later, in order to correct for the secondary spectrum, Peter Dollond began adding yet a third component to his lenses. One of these triplet lenses with a 3.75-inch (95-mm) aperture was purchased by Neville Maskelyne for the Royal Observatory at Greenwich, and another, of 3.8 inches (97 mm), was acquired by the wealthy connoisseur Dr. William Kitchener, who commented that "it shews the disc of the Moon and of Jupiter as white and as free from colour as a Reflector."[2] But progress in this direction was limited. Lenses larger than about 4 inches (10 cm) in diameter proved to be almost impossible to make because the flint disks always developed streaks and veins during casting.

SCHROETER MADE few observations of Mars after 1800. Instead, the most diligent observer of the period was a Frenchman, Honoré Flaugergues, who is best remembered today as the discoverer of the Great Comet of 1811. He had a private observatory at Viviers (Ardèche), in southeastern France, whose main instrument was a rather inefficient achromat of 44-foot (13.4-m) focal length, giving a usual magnification of only 90×. Flaugergues began observing Mars at the opposition of 1796, but carried out especially thorough studies at the oppositions of 1809 and 1813, the latter being the first perihelic opposition of the new century. He noted the more prominent patches on the disk, which appeared dark reddish to him, and calculated the times at which the planet's rotation ought to bring the same aspects into view again. He found great inconsistencies, however, and was unable to believe that he was actually observing the solid surface of the planet. Instead, he accepted Schroeter's conclusion that only atmospheric features were on view. "These patches seemed to me to be in general confused and badly defined," he wrote in 1809, "to an extent that it was difficult to distinguish exactly their outlines and their full extent. I can say only that normally the south part of the disk was the region of Mars which contained the principal patches."[3] As noted earlier, we now know that Mars is sometimes obscured by dust clouds and veils, and some historians claim that Flaugergues was the first to record them.[4] Here I must disagree; Flaugergues simply wasn't a good enough observer, and his telescope was too small and too crude to have produced trustworthy results. It is clear that for all his diligence, he achieved no significant advance over Schroeter or even Maraldi.

In addition to his rather confused observations of the spots on Mars, Flaugergues noticed the rapidity of the south polar cap's melting and remarked that if it consisted of ice and snow, "as everyone believes," this proved that Mars, despite its greater distance from the Sun, must be warmer than the Earth!

MEANWHILE, important developments had been taking place in optics. In 1799, a Swiss artisan named Pierre Louis Guinand had discovered that flawless disks as large as 6 inches (15 cm) in diameter could be cast from flint glass if the flint was stirred while it was cooling. Guinand moved to Munich in 1805, and there he joined forces with a brilliant young optician, Joseph Fraunhofer, who under Guinand's supervision became extremely proficient in glassmaking. Moreover, Fraunhofer practically devised the science of correctly designing achromatic objective lenses using only two components. By 1812 he had succeeded in producing a fine achromat 7.5 inches (19 cm) in diameter. Fraunhofer's refractors produced clear, brilliant images of the Moon and planets, and were a marked improvement over the large reflectors used by Herschel and Schroeter. By 1813, the year Schroeter's observatory at Lilienthal was destroyed by the French, the refractor was once more coming to dominate astronomy—indeed, the nineteenth century would become the "century of the refractor."

Among the first to use a Fraunhofer refractor to view Mars was Georg Karl Friedrich Kunowsky. He was a lawyer by profession, and served as Justizrat in Berlin, but Kunowsky was also a keen amateur astronomer. With a 4.5-inch (11-cm) Fraunhofer refractor, he made a number of sketches of the dark patches at the unfavorable opposition of 1821-22.[5] Unlike Schroeter and Flaugergues, however, Kunowsky came to the conclusion that the patches were fixed features of the surface. Nevertheless, his results were hardly definitive, and the question—163 years after Huygens had first sketched Syrtis Major—remained unresolved. That being the case, it is easy to understand why no one had seen any point in attempting to draw a map of the planet.

ALL OF THIS would change with the advent of Wilhelm Beer and Johann Heinrich Mädler, who began the next new era in the study of Mars—a period Flammarion described as the "geographical period." Thus far the study of Mars had advanced by slow and uncertain steps; after Beer and Mädler many uncertainties remained, but henceforth the results rested on a firmer basis. Flammarion, in his incomparable fashion, said it best: "Christopher Columbus was happy when he was halted by the American continent during his voyage of circumnavigation beyond Asia. Mars does not have its Christopher Columbus. He achieved fame by the single fact of touching America; a phalanx of astronomers has been busy for more than a century studying their celestial continent. But Beer and Mädler deserve to be remembered as the true pioneers in this new conquest."[6]

Mädler was born in Berlin in 1794, the son of a master tailor.[7] He had taught himself to read by the time he was four, and a maternal uncle, Paul Strobach, recognizing his ability, pushed for him to get a sound education. At age twelve he was sent to the Friedrich-Werdersche Gymnasium in Berlin. Meanwhile his interest in astronomy had surfaced, inspired by the comet Flaugergues had discovered—the Great Comet of 1811. Mädler proved to be an excellent scholar and seemed destined for an academic career, but then disaster struck—when he was nineteen, an outbreak of typhus claimed both his parents and his maternal uncle, and he found himself weighted with the responsibility of supporting four younger siblings. He abandoned his academic dreams and enrolled in the tuition-free Küsterschen Seminary in order to study to become an elementary school teacher. At the same time he began giving lessons as a private tutor, and in 1819 he founded a school in Berlin for children of parents with limited means—obviously not a highly lucrative way to make a living. Meanwhile, he began attending lectures at the University of Berlin and heard, among others, P. G. L. Dirichlet on higher mathematics and Johann Elert Bode and Johann Franz Encke on astronomy.

A turning point in his life came in 1824, when he met Wilhelm Beer, who had applied to him for private lessons in higher mathematics and astronomy. Beer, a well-to-do banker who had just taken over the family banking business from his father, Jakob Herz Beer, was then twenty-seven years old. His brothers were the poet Michael Beer and the composer Jakob Beer, who styled himself Meyerbeer and went on to become the most successful operatic composer of his day.

After he met Mädler, Beer decided to set up his own observatory. The main instrument was a 3.75-inch (95-mm) Fraunhofer refractor, which was mounted equatorially and equipped with a clock drive allowing it to follow the apparent drift of the stars. Beer set it up near his villa, on a platform in the famous Tiergarten under a rotatable dome 12 feet in diameter whose shutters opened upon a swath of sky 20° wide. The telescope was in place by 1828, and two years later Beer and Mädler began to use it for mapping the Moon, the work for which they are best remembered. (It has long been recognized that most of the actual mapping was done by Mädler; Beer's main contribution was in allowing him to use the observatory!)

They had been mapping the Moon for several months when Mars's perihelic opposition on September 19, 1830, presented them with a great opportunity. For several weeks around this date, Beer and Mädler observed Mars extensively. Their first goal was to precisely determine its rotation period, and at the same time they hoped to establish once and for all whether or not the patches on the surface of Mars were variable.[8]

Beer's 3.75-inch Fraunhofer refractor, though of modest size, was first class and gave a sharper definition than the larger reflectors of Herschel and Schroeter. Even so, Beer and Mädler found the Martian surface features to be weak and ill-defined—which explains how the sharp disagreement as to their nature and permanence could have gone on for so long. They wrote that

> the use of a micrometer did not seem convenient to us, the thickness of the threads causing more uncertainty in measurement of such fine objects than was produced by estimating by the eye

alone. The drawings were executed immediately at the telescope. Ordinarily some time elapsed before the indefinite mass of light resolved into an image with recognizable features. We next attempted to estimate the coordinates of the most distinct points, using the white spot at the south pole for the determination of the central meridian, and only then sketched in the remaining detail. . . . Finally, each of us compared the drawing with the telescopic image, so that everything shown was seen by both of us and hopefully may be considered fairly reliable.[9]

Their study left little doubt that the markings were constant. "Our observations," they wrote, "are thus in important disagreement with earlier ones. . . . The hypothesis, that the spots are similar to our clouds, appears to be entirely disproved."[10]

At the beginning of their observations, Beer and Mädler's attention had been struck by a small round patch "hanging from an undulating ribbon." This round patch had been represented very imperfectly by Herschel in 1783, and on several occasions by Schroeter in 1798, but Beer and Mädler were the first to show it clearly. It lay only 8° south of the equator, and they regarded it as a convenient reference point for determining the rotation period of Mars. Later astronomers concurred in the aptness of their choice—ever since Beer and Mädler, that feature has defined the zero meridian of Mars, and Camille Flammarion later named it the Meridian Bay (Sinus Meridiani). Rather than giving names to the various markings they mapped, Beer and Mädler simply designated them with letters—thus the small round patch was indicated by the letter *a,* Syrtis Major by *efh,* and so on. From their observations of the patch *a,* they put the rotation period at 24 hours, 37 minutes, 9.9 seconds.

In 1830, Beer and Mädler began a careful study of the south polar cap. They followed its rapid shrinking and noted that this continued until the Martian season corresponding to our mid-July. Then the cap began slowly to increase again. These observations lent strong support to the idea that it consisted of ice and snow.

Although Beer and Mädler were able to confirm some of their earlier results at the oppositions of 1832 and 1834-35, they added little that was new. In 1837 they began to use a much larger instrument, the 9.6-inch (24-cm) refractor of the Royal Observatory of Berlin. Despite this superior instrument, however, they were much hampered by Mars's greater distance from the Earth and the "almost unprecedented bad weather" in Berlin. Nevertheless, they were able to achieve some useful results. They revised the rotation period to 24 hours, 37 minutes, 23.7 seconds—very close to the presently accepted value. This was, they noted, 2 minutes shorter than the period published by William Herschel, but they were able to account for the discrepancy after carefully reviewing Herschel's records from 1777 and 1779. During the interval Herschel had used to compute the rotation period, Mars had completed one more rotation than he had realized. When this was factored in, the agreement with their own results was excellent.

In 1830, the rapidly shrinking south polar cap had been tilted toward the Earth; in 1837, it was the north polar cap. The two caps showed markedly different behavior; both were centered within a few degrees of the poles, but the south polar cap grew much larger. At the same time, its retreat was more rapid and complete; the smallest size Beer and Mädler recorded for the south polar cap was 6°, whereas the north polar cap never shrank below 12° or 14°.

There are perfectly logical reasons for these differences, and they have to do with the nature of the Martian seasons. To make this clear, it is useful to introduce a calendar based on L_s, the areocentric longitude of the Sun, which gives Mars's position in its orbit relative to the Sun-Mars line at the northern spring equinox, which marks the beginning of northern spring. This point is defined as $L_s = 0°$. The northern spring lasts from $L_s = 0°$ to $90°$, summer from $90°$ to $180°$, fall from $180°$ to $270°$, and winter from $270°$ to $360°$ (or $0°$). (As on Earth, seasons in the Martian southern hemisphere are $180°$ out of phase with those of the northern hemisphere, so that the southern hemisphere has its summer when the northern hemisphere has its winter, and vice versa.) Since Mars's year is almost twice as long as one Earth year—it lasts for 668.6 Martian days, or Sols (1 Sol = 24 hours, 37 minutes, 22.663 seconds)—it follows that Martian seasons will be much longer than those of Earth. They are also more unequal—a consequence of the much greater eccentricity of the Martian orbit. The details are given in table 1.

Since Mars's perihelion lies at $L_s = 250.87°$, the planet passes this point late in the southern hemisphere spring. Mars is then 26 million miles (43 million km) nearer to the Sun and receives 45 percent more solar radiation, than at aphelion ($L_s = 70.87°$). Thus south-

TABLE I. Duration of Martian Seasons Relative to Those of Earth

Areographic longitude (L_s)	Martian season		Duration on		
	Northern hemisphere	Southern hemisphere	Mars		Earth
			(Sols)	(days)	(days)
0–90°	spring	fall	194	199	92.9
90–180°	summer	winter	178	183	93.6
180–270°	fall	spring	143	147	89.7
270–0°	winter	summer	154	158	89.1

Source: C. M. Michaux and R. L. Newburn, *Mars Scientific Model.* Jet Propulsion Laboratory Document 606–1 (1972).

ern hemisphere springs and summers are shorter but much hotter than northern hemisphere springs and summers, with peak temperatures as much as 30°c higher. Conversely, the southern hemisphere autumns and winters, which occur near aphelion, are much colder and longer. The southern hemisphere is thus a place of extremes; the northern hemisphere is one of relative moderation.

The behavior of the polar caps reflects these idiosyncrasies. The south polar cap grows greatly during the southern hemisphere's long, cold winter and shrinks rapidly during the short, hot summer. The north cap, reflecting that hemisphere's more moderate seasons, does not vary between such wide extremes. There are compositional differences as well, about which I will have more to say later. The northern cap is mainly made up of water ice, the southern cap of frozen carbon dioxide.

In addition to their studies of the polar caps, Beer and Mädler continued to sketch the dark patches on the planet. The dark area surrounding the north polar cap seemed to undergo especially marked changes—in 1837 it was of unequal width and not everywhere equally black, though it was still noticeably darker than the other spots; by 1839 it had become faint and narrow. These changes would be explained, they suggested, if the dark area was marshy soil moistened by meltwater from the retreating snow. There were times in 1837 when Mars was nearly featureless apart from the polar patch, which always remained distinctly in view. Indeed, even the patch *efh* (Syrtis Major) was by no means well defined. Despite the unsteadiness of the atmosphere at Berlin that year, it is tempting to believe that some of the obscurations may have been genuinely Martian—caused by the dust clouds and veils that are known to develop from time to time.

In 1840, Mädler combined all the observations and drew the first map of Mars ever made. Admittedly, it leaves much to be desired, but it nevertheless represents a tremendous step forward. That same year, Mädler left Berlin to become director of the Dorpat Observatory in Estonia, and at the opposition of 1841 he made only a few observations with the 9.6-inch (24-cm) Dorpat refractor. The experience of 1837 was repeated; though he was able to recover some of the spots of earlier years, he looked in vain for others, including the prominent round patch *a*. He was no longer so certain of the long-term stability of the markings.

BEER AND MÄDLER tower so far above their contemporaries that there is a distinct danger of forgetting the other observers who were active at that time. There was William Herschel's son John, for example, a great astronomer in his own right. He considered the ocher areas of Mars to be continents and suggested that they might be similar to the red limestones of Earth; the dark areas he regarded as seas since, he noted, water absorbs light more strongly than land. He also recorded greenish tints in the seas, though these he suspected of being illusory—a mere optical effect resulting from contrast with the ocher areas. As we shall see, the colors of Mars would become one of the most debated features of the planet. Another leading astronomer of the period was François Arago, who in 1830 became director of the Paris Observatory. He too remarked on the colors of the planet, finding a rosy tint obvious at low powers; with larger instruments, however, this passed successively to orange, yellow, and finally to lemon.[11] He agreed with the younger Herschel that the greenish tint of the dark areas was a contrast effect.

The next perihelic opposition took place on August 18, 1845, and it is memorable for the discovery by Ormsby MacKnight Mitchel, of the Cincinnati Observatory, of the large detached area of the south polar cap (centered at 75° s, 320° w) that begins to separate from the cap's rim at about the same Martian seasonal date each year (L_s about 240°). The final remnants do not disappear for another twenty or thirty days. This feature is still sometimes referred to as the "Mountains of Mitchel," but the name is a misnomer—instead of being mountainous, the region is actually depressed relative to the surrounding terrain.

As BETTER telescopes became more widely available, and more and more people acquired the passion to observe, the number of observations of Mars increased. The improvement in the quality of drawings in the thirty years after Beer and Mädler did their work is dramatic. In 1856, for example, the English amateur astronomer and pioneer photographer Warren De la Rue made several excellent drawings with a 13-inch (33-cm) reflector—in one, Syrtis Major, or the Hourglass Sea, appears very narrow; in another, the conspicuous round patch described by Beer and Mädler is represented as a pointed tongue.

At the opposition of 1858, Mars was extensively observed by the Jesuit astronomer Angelo Secchi, director of the observatory of the Collegio Romano in Rome.[12] Secchi used a 9.5-inch (24-cm) equatorial refractor with magnifying powers of 300--400×. In one of his first observations, on May 7, 1858, he described "a large triangular patch, blue in color." This was none other than the well-known Hourglass Sea, but Secchi gave it a different name: he called it the "Atlantic Canale," commenting that it "seems to play the role of the Atlantic which, on Earth, separates the Old Continent from the New." Thus the first occurrence of the fateful term *canale*, which in Italian can mean either "channel" or "canal." Secchi himself was inconsistent; later he called the same feature the "Scorpion"—a not inapt comparison to its appearance at the time.

Secchi was impressed with the great variety in the tints of the Martian features and even attempted to make the first color representations of the planet in pastels. He described the dark areas surrounding the polar caps as "ashen colored," but most of the other dark areas appeared bluish, with an occasional tint of green. As for the nature of the Martian markings, he wrote:

It is clear that the variations [in the polar caps] can be explained only by a melting of the snow or a disappearance of the clouds covering the polar regions. These aspects also prove that liquid water and seas exist on Mars; this is a natural result of the behavior of the snows. This conclusion is confirmed by the fact that the blue markings which we see in the equatorial regions do not change sensibly in form, whereas the white fields in the neighborhood of the poles are adjacent to reddish fields which can only be continents. Thus, the existence of seas and continents . . . has been today conclusively proved.[13]

At the perihelic opposition of July 1860, Mars was very far to the south, and thus difficult to observe from the Northern Hemisphere of the Earth, where most observatories were located. At the next opposition, in 1862, Mars was more favorably placed, and a number of observers took advantage of the opportunity, including Secchi in Rome, Lord Rosse in Ireland, and William Lassell on Malta. An important series of drawings was made by the director of the Leyden Observatory, Frederik Kaiser, who reworked the rotation period by comparing his drawings with those made by Hooke and Huygens in the seventeenth century—his result was 24 hours, 37 minutes, 22.6 seconds.[14] He also compiled the best map of the planet up to that time, continuing to use letters to designate the various features. Another skillful observer was J. Norman Lockyer, an English astronomer destined to play a prominent role in late Victorian science.[15] Lockyer used a 6.25-inch (16-cm) refractor, and his drawings of Mars, in the estimation of E. M. Antoniadi, "gave us the first really truthful representation of the planet." [16] This English astronomer accepted the basic permanence of the dark areas, though he noted that there were obvious variations over time. Thus, to give but one example, Solis Lacus—the "Oculus," or Eye, as it was then known—which had been depicted as nearly circular by Beer and Mädler, had become distinctly elongated by Lockyer's time. Still other variations Lockyer believed to be accounted for by clouds—indeed, his disks indicate that there were some rather persistent veils over Mare Erythraeum in 1862. Like Secchi, he believed that the greenish areas were seas and the ruddy areas continents.

This belief was by now generally accepted, though Secchi was premature, to say the least, in considering that it had been proved. A few astronomers, at any rate, remained skeptical. An Oxford professor of geology named John Phillips, one of the most active observers of Mars at the opposition of 1862, wrote that "a great part of the northern area appeared bright, and often reddish, as [if] it were land, while a great part of the southern area was of the grey hue which is considered to indicate water, but relieved by various tracts of a tint more or less approaching to that of the brighter spaces of the northern hemisphere."[17] Though he hedged regarding whether the dark patches were actual seas or mere gray plains like those of the Moon, he pointed out that if they *were* actual seas, the Sun's specular reflection ought to be visible on them. According to later calculations by G. V. Schiaparelli, this reflection would appear as bright as a star of the third magnitude. For many years the precise spot where the starlike image should be sought was published in physical ephemerides of the planet, but it was never seen. In the meantime, a different explanation of the Martian surface features was published by Emmanuel Liais, a French astronomer who left the Paris Observatory and moved to Brazil, where he became director of the observatory of Rio de Janeiro. Liais proposed that the ruddy areas were deserts of sand, and the dark patches vast tracts of vegetation, although it must be admitted that his ideas received little attention at the time.[18]

THE OPPOSITION of December 1, 1864, was not quite as good as that of 1862—Mars attained a maximum diameter of only 17.3″ —but the opposition was nevertheless memorable for the study carried out by Rev. William Rutter Dawes, the son of a mathematics teacher and once an astronomer on an expedition to Botany Bay, Australia. Dawes had studied medicine as a young man and later became a clergyman with a small Independent congregation at Ormskirk, north of Liverpool. After failing health forced him to give up his congregation, he devoted himself entirely to astronomy. In the 1840s he was an assistant at the private observatory of a wealthy businessman, George Bishop, at St. John's Wood, London. After his second marriage—to an Ormskirk solicitor's widow— Dawes acquired the financial independence he needed to set up his own private observatories, first at Cranbrook, Kent, and then, from 1857 until his death in 1867, at Haddenham, Buckinghamshire. He was an exceptional observer noted for the keenness of his sight; but eagle-eyed as he was at the telescope, he was so terribly nearsighted that he could pass his wife in the street without recognizing her!

Dawes had already made some drawings of Mars in 1862 and at earlier oppositions. In 1864, he used an 8-inch (20-cm) Cooke refractor, usually with a magnifying power of 258×. His drawings, wrote Richard Anthony Proctor, "are far better than any others. . . . The views by Beer and Mädler are good, as are some of Secchi's (though they appear badly drawn). Nasmyth's and Phillips', De La Rue's two views are also admirable; and Lockyer has given a better set of views than any of the others. But there is an amount of detail in Mr. Dawes' views which renders them superior to any yet taken."[19] Camille Flammarion concurred: "The drawings by . . . Dawes brought a new precision to studies of Mars."[20] A case in point: what Beer and Mädler had taken as a small, perfectly round spot (the feature they had designated a) was seen by Warren De la Rue as pointed and by Lockyer as an elongated patch; Dawes, however, resolved it into a bay with two forks, whose extensions, he noted, gave the impression of "two very wide mouths of a river, which however I could never trace. . . . It may be that the sea has receded from that part of the coast, and left the tongue of land exposed."[21] This was the famous "Dawes' forked bay"—a name that is still used from time to time.

By the 1860s, Beer and Mädler's chart was hopelessly out of date. Kaiser's was a definite improvement, but he had done nothing to improve on the old lettering system of nomenclature, which was proving more and more inconvenient. Over the years, a few names had come into general but unofficial use for the most prominent or singular features—Hourglass Sea and Oculus, for example—but most of the features remained unnamed.

The first attempt to find a more suitable Martian nomenclature was made by Proctor. He was a prolific writer of popular books on astronomy and, as we have seen, a great admirer of Dawes. In 1867, Proctor drew up a map of Mars based, somewhat crudely, on Dawes's drawings.[22] He explained his system of nomenclature by saying, "I have applied to the different features the names of those observers who have studied the physical peculiarities presented by Mars." For later reference, some of his names are here paired with those later proposed by Schiaparelli:

Kaiser Sea	Syrtis Major
Lockyer Land	Hellas
Main Sea	Lacus Moeris
Herschel II Strait	Sinus Sabaeus

Dawes Continent	Aeria and Arabia
De La Rue Ocean	Mare Erythraeum
Lockyer Sea	Solis Lacus
Dawes Sea	Tithonius Lacus
Madler Continent	Chryse, Ophir, Tharsis
Maraldi Sea	Mares Sirenum and Cimmerium
Secchi Continent	Memnonia
Hooke Sea	Mare Tyrrhenum
Cassini Land	Ausonia
Herschel I Continent	Zephyria, Aeolis, Aethiopis
Hind Land	Libya

Proctor's nomenclature has often been criticized, mainly because so many of his names honored English astronomers, but also because he used many names more than once—in particular, Dawes appeared no fewer than six times (Dawes Ocean, Dawes Continent, Dawes Sea, Dawes Strait, Dawes Isle, and Dawes Forked Bay). Moreover, as Schiaparelli later complained, Proctor's map did "not even give an accurate representation of the observations of Dawes himself."[23] Even so, Proctor's names are not without charm, and for all their shortcomings they were a foundation on which later astronomers could—and did—improve.

THROUGH DAWES'S skillful work, the main outlines of the Martian "seas" and "continents"—as most of the astronomers of the day assumed them to be—had been reliably depicted. Moreover, on those same drawings on which the broader features of the Martian disk were so artfully and accurately portrayed, tentative indications of a class of finer details were now rising for the first time above the threshold of perception. From some, though not all, of the pointed extensions of the dark seas there seemed to be prolongations into thin, wispy streaks, which imperceptibly dissolved again into the broad ocher parts of the planet. The same ocher areas, Secchi had noted, were not uniform but seemed to be filled with fine detail, whose nature "it was impossible to depict, or even for the imagination to capture."[24]

All of this was suggestive. The broader features of the Martian surface may have been completely defined, but there was something more remaining to be discovered—a fine print, in relation to which humankind stood in the 1860s where Christiaan Huygens had stood in relation to the big print two centuries earlier.

The observations of the early pioneers had been limited by the chromatic aberration and diffraction that afflicted their instruments. Diffraction is a consequence of the wave nature of light. Because of it, a telescope can never form an image of a star as a perfect point; instead, the image is a small disk surrounded by a series of rings—the larger the telescope, the smaller the apparent disk and the more closely spaced the rings. In the case of a double star, if the diffraction patterns overlap too much, the components will remain unresolved. (Dawes himself is perhaps best remembered today for working out what is known as Dawes's limit: the aperture of a telescope, determined empirically, needed to just separate the components of close double stars.) Diffraction, of course, enters just as critically into observations of planetary surface detail. A thin line on a planet, for example, becomes widened by diffraction into a band of decreased intensity on both sides. If a line is less than a certain width, its contrast with the background is so much reduced that the eye is simply unable to grasp the weakened tones.

The limits to visibility set by diffraction are invincible; the only remedy is to increase the aperture width of the telescope being used. It is this increase in aperture or the corresponding reduction in the diffraction limit that explain why Beer and Mädler, with a 3.75-inch (95-mm) aperture, were able to make out only a round spot while Dawes, with 8 inches (20 cm), saw a forked bay.

Secchi, Kaiser, Lockyer, and Dawes, the best observers of their day, all used instruments with apertures of less than 10 inches (25 cm). With such instruments it was possible to reach, during the steadiest moments of the atmosphere of Earth, the limits of visibility set by diffraction. Naturally, superior results were expected from much larger instruments. In 1862, the 18.5-inch (47-cm) Clark refractor of the Dearborn Observatory, in Chicago, had just come into service, surpassing the 15-inch (38-cm) refractors of the Harvard and Pulkova observatories, which had been built in the early 1840s. But the era of great telescopes was only beginning. By 1870, the 25-inch (63-cm) Cooke refractor had been set up on the estate of wealthy amateur R. S. Newall at Gateshead, in northern England;

and in 1873 the 26-inch (66-cm) Clark went into operation at the U.S. Naval Observatory in Washington, D.C. Unfortunately, the locations for these large instruments could hardly have been more ill-chosen, and it soon became painfully apparent that there are other waves no less pertinent to planetary observation than light waves: atmospheric waves, which impose their own barrier to seeing and are an ever-present foil to attempts to reveal finer planetary surface markings from the Earth. Since a larger telescope takes in larger and stronger areas of atmospheric turbulence, beyond a certain point—about 12 or 16 inches (30 or 40 cm)—the advantages over smaller instruments will be partially or entirely offset by blurring, at least much of the time. It follows that a smaller instrument used in good conditions can surpass the results of a much larger one used in terrible seeing (bad atmospheric conditions). After fifteen frustrating years of trying to use his instrument, Newall wrote, "Atmosphere has an immense deal to do with definition. I have had only one fine night since 1870! I then saw what I have never seen since."[25]

THUS FAR the story of Martian exploration has been intertwined with the optical improvement of the telescope. Now we enter a new era, in which the pursuit of the best seeing—the quality of the air, which becomes best when it arranges itself into steady, stable layers above the ground—becomes equally important to the quest for Martian detail.

1877

On September 5, 1877, Mars came to a perihelic opposition in the constellation Aquarius, approaching to within 35 million miles (56 million km) of the Earth. Since the last perihelic opposition in 1860, the 26-inch (66-cm) refractor of the U.S. Naval Observatory, located at Foggy Bottom on the banks of the Potomac River in Washington, D.C., had gone into operation. In 1877, Asaph Hall was in charge of it and planned to use it to search for Martian satellites (fig. 4).

Hall was the son of a failed clock maker. He and his wife, Angelina, were working as schoolteachers in Shalersville, Ohio, in the 1850s when Hall decided that he wanted to become an astronomer. Though he had not received much formal training (he had stayed only a year at the University of Michigan at Ann Arbor before leaving because of a lack of funds), he applied to become an assistant at the Harvard College Observatory. William Cranch Bond, the observatory's director, was, like Hall, the son of a clock maker. He too had begun his astronomical career without many advantages, and he duly hired the young man to assist himself and his son, George Phillips Bond.

The position was not a lucrative one, and Hall later recalled that when he first met G. P. Bond, who had been away from the observatory when Hall arrived, Bond "had a free talk with me, and found out that I had a wife, $25 in cash, and a salary of $3 a week. He told me very frankly that he thought I had better quit astronomy, for he felt sure that I would starve. I laughed at this, and told him my wife

FIGURE 4. Asaph Hall in 1871, six years before he discovered the Martian satellites. (Courtesy Mary Lea Shane Archives of the Lick Observatory)

and I had made up our minds that we were used to sailing close to the wind, and felt sure we would pull through."[1]

Hall left Harvard in 1863 for the U.S. Naval Observatory in Washington, D.C., and took charge of its great refractor, the first of the great refractors made by miniature painter–turned-optician Alvan Clark, in 1875. For two years it had been in the hands of Simon Newcomb, who was more interested in mathematical astronomy than in observing, and his assistant, Edward Singleton Holden. Hall later recalled finding "in a drawer in the Eq[uatorial] room a lot of photographs of the planet Mars in 1875. From the handwriting of dates and notes probably Holden directed the photographer, but whoever did the pointing of the telescope had . . . satellites under his eye."[2]

Hall later retraced the steps that led him to undertake his own search for Martian satellites:

> In December, 1876, while observing the satellites of Saturn I noticed a white spot on the ball of the planet, and the observations of this spot gave me the means of determining the time of the rotation of Saturn, or the length of Saturn's day, with considerable accuracy. This was a simple matter, but the resulting time of rotation was nearly a quarter of an hour different from what is generally given in our text books on astronomy: and this discordance, since the error was multiplied by the number of rotations and the ephemeris soon became utterly wrong, set before me in a clearer light than ever before the careless manner in which books are made, showed the necessity of consulting original papers, and made me ready to doubt the assertion one reads so often in the books, "Mars has no moon."[3]

On looking further into the matter, Hall learned that William Herschel had looked unsuccessfully for satellites in 1783, and that the director of the Copenhagen Observatory, H. L. d'Arrest, had done so in 1862 and 1864. (He did not mention Holden's photographic search with the great Washington refractor.) Of these searches, d'Arrest's had been the most thorough. He had been guided by rough calculations of the distance from the planet at which a satellite could exist before it was wrenched away by the Sun into its own planetary orbit, and had set this limit at a distance corresponding to 70′ of arc from the planet at greatest elongation.

Hall, on redoing the calculation, realized that the actual limit ought to be more like 30' of arc, and that Martian satellites were likely to be found even closer than that to the planet. He began to suspect, therefore, that d'Arrest, for all his thoroughness, had not paid sufficient attention to the inner space near the planet.[4]

When Hall began his quest in early August, he naturally wanted to work alone, so as to receive full credit in the event of a discovery. By great good luck, Holden, his assistant, was invited by Henry Draper to Dobbs Ferry, New York, "at the very nick of time."[5] Hall began by scrutinizing faint stars at some distance from Mars itself, but each one soon dropped behind the planet, proving it to be an ordinary field star.[6] Next he pressed the search closer, "within the glare of light that surrounded [Mars]," using special observing techniques to reduce the glare, such as "sliding the eyepiece so as to keep the planet just outside the field of view, and then turning the eyepiece in order to pass completely around the planet." On the night of August 10, the first on which Hall attempted to examine the inner space near Mars, he found nothing, but the seeing on the banks of the Potomac was horrible that night, and the image of the planet appeared "very blazing and unsteady." He was on the verge of giving up, but Angelina encouraged him to have one more try, and the next night, at half past two, he found a suspicious object which he referred to in his notebook only as "a faint star near Mars." He scarcely had time to secure its position before the fog began rolling in from the Potomac. The next few nights were cloudy. On August 15, the sky cleared at eleven o'clock, but the atmosphere, Hall noted, was still "in a very bad condition." Not until August 16 did he again find the "star near Mars," which proved, in fact, to be the outer satellite. That night he showed the object to another assistant, George Anderson, but told Anderson to "keep quiet" about it. On August 17, while waiting for that satellite to reappear, he discovered the inner one. In closing his observing notes for the night, he remarked: "Both the above objects faint but distinctly seen both by G. Anderson and myself." Hall had by this time "spilled the beans" to Simon Newcomb, and on August 18, Hall and Anderson were joined in the dome by David Peck Todd, Newcomb, and William Harkness. Todd noted: "Seeing extremely bad: still I saw the companion without any difficulty. 'Halo' around the planet very bright, and the satellite was visible in this halo." Only then did Hall announce the discovery of the two satellites. Newcomb tried to gain a share of the credit for himself, implying in an article that appeared in the *New York Tribune* two days after the discovery was announced that Hall had not fully appreciated what he had found until he—Newcomb—had worked out the period of revolution from the preliminary observations.

Meanwhile, in New York, Holden and Draper were also getting into the act. On August 28, Holden announced that they had used Draper's 28-inch (71-cm) reflector to discover a third satellite, and on returning to Washington, Holden claimed to have found yet a fourth. Hall was skeptical and wrote to Arthur Searle of Harvard: "I think it will turn out that the Draper-Holden moon and the recent Holden moon do not exist."[7] He attempted to confirm these alleged discoveries with the Washington refractor without success, and later computations showed that Holden's moon did not even obey Kepler's laws of motion. "Its existence was therefore a mathematical impossibility," Hall wrote to Edward C. Pickering of Harvard Observatory, adding bitterly: "If I were to go through this experience again other people would verify their own moons."[8]

Rumors of Holden's spurious moons would continue to circulate in the astronomical community for years, and Holden became known as the man "who had set all Washington astronomers laughing by detecting a . . . satellite of Mars with an impossible period and distance, and remaining deceived by it for months!"[9] But Holden, in Hall's view, at least, had behaved more admirably than Newcomb. As late as 1904 Hall was still bitter about Newcomb's attempt to usurp credit for the discovery of the satellites, and wrote to S. C. Chandler, Jr.: "Newcomb was greatly excited over my discovery. Holden was away, and Draper made a blunder, and afterwards Holden behaved very well. Newcomb felt disappointed and sore, and something is to be allowed for human nature under such circumstances. He was always greedy for money and glory."[10]

In response to a suggestion by Henry Madan of Eton, England, Hall named the satellites Phobos (Fear) and Deimos (Flight), after the attendants of Mars mentioned in the fifteenth book of Homer's *Iliad*: "He spake, and summoned Fear and Flight to yoke his steeds." Hall continued to watch the new satellites until the end of October, and his observations gave him the information he needed to work out the mass of Mars—the amount of matter it contains—from its effect on the moons' motions. It was 0.1076 times that of the Earth (a value very close to the currently accepted value of 0.1074). We will examine the satellites in greater detail in chapter 14.

Phobos and Deimos were seen not only by Hall but by viewers using much smaller instruments—indeed, Deimos was glimpsed by Hall, John Eastman, and Henry M. Paul with the U.S. Naval Observatory's own 9.6-inch (24-cm) refractor. This only goes to show that the discovery of the satellites of Mars owed quite as much to Hall's insight—his imagination and willingness to doubt conventional wisdom—as to the size of his glass. As he later wrote, "All that was needed was the right way of looking, and that was to get rid of the dazzling light of the planet."[11] He was confident that with the right way of looking, the satellites could have been found "very easily" even with Harvard's 15-inch refractor in 1862.

THERE WERE other important studies of Mars in 1877 as well. Nathaniel Green, an amateur astronomer and professional portrait artist who at one time had given lessons in painting to Queen Victoria, made a careful study with his 13-inch (33-cm) Newtonian reflector on the eastern Atlantic island of Madeira, a site renowned for its pellucid skies. Green drew up a map and also noted brightenings at the limb and terminator which he identified correctly as morning and evening clouds (fig. 5).[12] Another valuable set of observations was made by Englishman Henry Pratt, using an 8-inch (21-cm) reflector. Pratt reported that "in moments of the finest definition the markings have exhibited a *stippled* rather than a streaked character, and glimpses were obtained of a structure so complicated and delicate that the pencil cannot reproduce it. . . . Frequently what at first sight appeared as a broad hazy streak has been, by patient watching for the best moments, resolved into several separate masses of shading enclosing lighter portions full of very delicate markings."[13]

But by far the most important development of the memorable 1877 opposition—after Hall's discovery of the satellites, of course—was the landmark study of the planet begun by the Italian astronomer Giovanni Virginio Schiaparelli, who would become the leading expert on the planet for the next two decades (fig. 6).

Schiaparelli was born on March 14, 1835, in the town of Savigliano, in the Piedmont region of northwestern Italy, not far from the French border.[14] The town lies among the foothills of the Alps

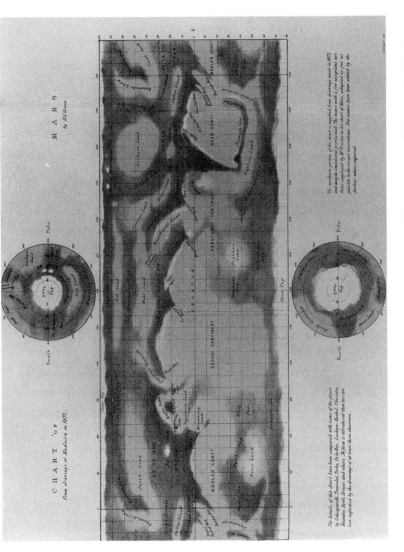

FIGURE 5. Nathaniel Green's 1877 map. (From N. E. Green, "Observations of Mars, at Madeira in Aug. and Sept., 1877," *Memoirs of the Royal Astronomical Society* 44 [1877–79], fol. p. 138)

and is overlooked by an ancient Benedictine abbey. His interest in astronomy was first awakened when his father, a furnace maker, took the four-year-old Schiaparelli outside on a clear and serene night and pointed out some of the constellations. "Thus, as an infant," Schiaparelli later wrote, "I came to know the Pleiades, the Little Wagon, the Great Wagon. . . . Also I saw the trail of a falling star; and another; and another. When I asked what they were, my father answered that this was something the Creator alone knew. Thus arose a secret and confused feeling of immense and awesome things. Already then, as later, my imagination was strongly stirred by thoughts of the vastness of space and time."[15] Young Schiaparelli's interest was further stimulated by the total eclipse of the Sun of July 8, 1842, which he observed with his mother through the window of the family's *casa,* and by the instruction of a learned priest of Savigliano, Paolo Dovo, who lent him books and gave him, from the campanile of the church of Santa Maria della Pieve, his first views through a small telescope of the phases of Venus, the moons of Jupiter, and the rings of Saturn.

On completing the rather rudimentary course that was all the elementary schools of Savigliano had to offer, Schiaparelli went on to the University of Turin, from which he graduated in 1854 with a degree in architectural and hydraulic engineering. For a time he employed himself in the private study of astronomy, mathematics, and languages, and in November 1856 he received an appointment as a teacher of mathematics in an elementary school in Turin. But his heart was not in it. Instead, as he later wrote, "without taking into account my almost absolute poverty, I formed the project of devoting myself to astronomy, which was not done without much opposition on the part of my parents."[16]

His great opportunity came in February 1857 when the Piedmontese government awarded him a small stipend, enabling him to receive training in astronomy. He spent two years studying at the Royal Observatory in Berlin, which was then under the directorship of Johann Franz Encke, and another year at the Pulkova Observatory working under Wilhelm Struve. Then, in 1860, he returned to Italy to take up a post as *secondo astronomo* under Francesco Carlini at the observatory at the Brera Palace in Milan, which had been founded by the famous Jesuit astronomer Roger Boscovitch a century earlier. The instruments at the observatory were an-

FIGURE 6. Giovanni Virginio Schiaparelli.
(Courtesy Luigi Prestinenza)

tiquated and woefully inadequate; there was only an old equatorial
sector and a meridian circle with lenses of 4 inches (10 cm) aperture. Nevertheless, Schiaparelli made the best use of the modest re-
sources available to him, and in April 1861 employed the equatorial
sector to discover the sixty-ninth asteroid, Hesperia. The following
year, Carlini died and Schiaparelli succeeded him as director.

The 1860s saw immortal work in Milan: the brilliant investigation in which Schiaparelli showed that the August meteors follow
the same orbit as that of the bright comet Swift-Tuttle (1862 III),
thereby forging a link between comets and meteors and at the same
time answering the question about falling stars that he had put to
his father so long before.[17] For this work he was awarded the prestigious Lalande Prize of the French Académie des Sciences in 1868.

The fame of his meteor work and the growing national pride

FIGURE 7. The 8.6-inch Merz refractor of the Brera Observatory. This is the instrument with which Schiaparelli discovered the *canali;* he used it again at the oppositions of 1877, 1879, 1882, and 1884. In 1886, 1888, and 1890, his principal instrument was a 19-inch refractor. (Courtesy Brera Observatory, Milan)

of the recently organized Kingdom of Italy brought Schiaparelli a more powerful telescope, an 8.6-inch (22-cm) refractor made by Merz, Fraunhofer's successor, which was installed on the roof of the Brera Palace in 1874 (fig. 7). At first, Schiaparelli used it mainly to measure double stars. Over the next twenty-five years he would make 11,000 such measurements, and the keenness of his eyesight is attested by the close separations of some of the double stars on which he successfully put the wires of the micrometer (incidentally, Schiaparelli, like Dawes, was very nearsighted; again, this did not interfere with his work at the eyepiece). Robert G. Aitken, a highly respected observer of double stars with the much larger telescope at the Lick Observatory, would later say of Schiaparelli's measures of the double star β Delphini that "the residuals shown in Schiaparelli's measures . . . will not seem very large—in fact it is surprising that measures of such a pair could be obtained at all with so small a telescope."[18]

In 1877, Schiaparelli began his studies of the planets. Before describing what he found out about Mars, I will summarize some of the results he obtained on the other planets. At the time, the rotations of all the planets from Mercury to Mars were believed to be about twenty-four hours. The rotation of Mars had, of course, been established beyond dispute, and indeed was well known to within a tenth of a second. However, Schiaparelli had little confidence in the rotations ascribed to Mercury and Venus, and he decided to investigate the matter further.

Venus is brilliant but difficult to observe because it usually shows only nebulous and ill-defined surface markings. In December 1877, Schiaparelli made out a pair of bright oval spots near the southern cusp of the planet, and also a shadowy streak. The markings were unusually conspicuous by Venusian standards, and he kept them under observation for two months, during which he failed to detect the slightest change in either their form or their position relative to the terminator. He therefore concluded that Venus's rotation was very slow, between six and nine months, and probably equal to the period of its revolution—224.7 Earth days.[19] This announcement received support from some quarters, and disagreement from others. Indeed, the question of Venus's rotation continued to vex visual observers right up to the early 1960s. It was finally settled only by advanced methods of radio astronomy.[20]

Unlike brilliant Venus, Mercury had received little attention from astronomers; being innermost to the Sun, it is notoriously difficult to observe. Schroeter, in 1800, had offered the only positive determination; he had observed exclusively in twilight periods, and he deduced from the blunted appearance of the southern cusp, which seemed to be unchanged from one night to the next, that the period of rotation must be right around twenty-four hours. Schiaparelli found the twilight conditions generally unfavorable because of the planet's low altitude, and instead decided to try to observe Mercury during broad daylight, when it was higher in the sky. He made the first tests in June 1881, and was encouraged enough to plan a regular study in 1882. That year he made every effort to keep the planet under continuous surveillance, observing it on February 4–10, March 31–April 28, May 24–31, August 5–21, and September 19–30. Although most of the observations were made around the times of Mercury's greatest elongations from the Sun, the planet was near superior conjunction in August, and Schiaparelli succeeded in following the small gibbous disk, then only 4″ of arc across, to within 3.5° of the Sun. With his fine refractor he made out markings on the surface which, he wrote, usually appeared "in the form of extremely faint streaks, which under the usual conditions of observations can be made out only with the greatest effort and attention."[21]

Schiaparelli's best observations were made in 1883–84. He was on the verge of making his results public then, but he decided to wait until he had had a chance to observe Mercury with the observatory's new 19-inch (49-cm) refractor, which was installed in 1886. These studies added nothing substantially new. Finally, in 1889, he was ready to announce his main result: the rotation period of Mercury, he declared, is equal to that of its revolution, eighty-eight days, and thus one side of the planet is always perpetually in daylight and the other in darkness. Though his observations had indicated a definite slow drift of markings across the disk, he explained this as an effect of libration, already well known in the case of the Moon, with its captured rotation with respect to the Earth. The Moon's libration is simply a result of the fact that the constant rate of the Moon's axial spin gets out of step with its changing velocity in its elliptical orbit, thereby producing an apparent rocking movement to and fro. Mercury, because of its very eccentric orbit around the Sun, would be expected to have a very marked libration, amounting to some 47° 21′ in longitude. But even granting such a broad allowance for variation in the observed position of the planet's features, Schiaparelli still found discrepancies, and he remarked to the English astronomer William F. Denning that the markings were "extremely variable." Sometimes they seemed to be "partially or totally obscured." Moreover, the planet showed "some brilliant spots which change their position." Thus Schiaparelli concluded that the surface of Mercury was sometimes covered by "veils [of] . . . more or less opaque condensations produced in the atmosphere of Mercury, which from afar presents aspects analogous to those which our Earth would show from a similar distance."[22]

Other observers, including Henri Perrotin, Percival Lowell, and René Jarry-Desloges, confirmed the eighty-eight-day rotation period. So, most notably, did the skilled Greco-French astronomer Eugène Michael Antoniadi, who between 1924 and 1929 carried out a careful study of Mercury with the 33-inch (83-cm) refractor at Meudon Observatory, near Paris. He agreed with Schiaparelli's rotation period, and also supported the existence of Mercurial clouds. Among the leading observers of the planet, only the French astronomer Georges Fournier was unable to reach a definite conclusion as to the rotation of Mercury.[23]

In fact, Fournier's diffidence proved to be justified. The true rotation of Mercury was discovered only in 1965 by radio astronomers. Instead of being equal to the period of revolution, 88 days, it is precisely two-thirds of this—58.65 days. How could the visual observers have been so far wrong? It turns out that 58.65 days is not only two-thirds of Mercury's period of revolution around the Sun, it is close to half of the period between its successive appearances in the same phase as viewed from the Earth (the synodic period, which in Mercury's case is 116 days). Thus, when Mercury comes to its greatest elongations from the Sun and is best placed for observation, astronomers in the Northern Hemisphere tend to see the same features; it is natural to conclude from this that the planet always keeps the same face toward the Sun.[24] The selective observation of the planet through periodic observing windows has been called the stroboscope effect. The synchronism is not perfect; after about seven years new regions of the planet begin to swing into view—but by the time seven years had passed, Schiaparelli had already made up his mind and had given up regular observing of the planet. As for his successors, their results demonstrate only too clearly that once a definite expectation is established, it is inevitable that subsequent observers will see what they expect to see, refining their expectations in a continuing process until finally everyone sees an exact and detailed—but ultimately fictitious—picture.

Even Antoniadi was misled; ironically, by the time he began his study of Mercury, the same features had returned to the disk that had been present for Schiaparelli's observations. He too kept the planet under observation for only seven years, and thus it is not surprising that (with due allowance made for differences in their drawing styles) Antoniadi's chart should appear almost identical with Schiaparelli's, or that the drawings of later observers appear to have been, in the words of Clark Chapman and Dale Cruikshank, "subconscious reproductions of Antoniadi's chart."[25]

I HOPE the reader will forgive me for going into the episode of Mercury's rotation at such great length. It provides, I think, an excellent introduction to the Martian "canal" affair because it clearly illustrates what Antoniadi once referred to as "the snares awaiting the observer at each stage of his work,"[26] snares from which no observer, no matter how skillful or cautious, can ever be entirely immune.

These snares are nowhere better documented than in the history of the famous, or infamous, Martian canals. We now return, therefore, to the fall of 1877, with Mars close to the Earth. Schiaparelli was eager to test the 8.6-inch Merz refractor—which he had been using for two years to study double stars—on Mars. It is clear that he did not at first intend to devote himself to a continued series of observations of the planet. Rather, he explained,

> I desired only to experiment to see whether our refractor . . . possessed the necessary optical qualities to allow for the study of the surfaces of the planets. I desired also to verify for myself what was said in books of descriptive astronomy about the surface of Mars, its spots, and its atmosphere. I must confess that, on comparing the aspects of the planet with the maps that had been most recently published, my first attempt did not seem very encouraging.[27]

Further study showed, however, that his drawings agreed quite well with the best made at previous oppositions, such as those by Kaiser and Lockyer. Thus, on September 12, 1877, he resolved on a careful study for the purpose of drawing up a new map of the planet. He generally used a magnifying power of 322× on the Merz refractor (and later, when the disk had become very small as the planet receded to a great distance from the Earth, 468×). Meticulous observer that he was, he was not satisfied to rely on eye estimates of the positions of features. Instead, he based his map on micrometric measures of the longitudes and latitudes of sixty-two distinctly recognizable points on the planet. The resulting map was a tremendous advance over anything that had appeared before. It was, Camille Flammarion declared, "a truly remarkable piece of work, and one showing features which the old observers of Mars could never have suspected. It depended for its successful completion on an unflagging persistence, an excellent eye, a rigorous method of observation and a good instrument." [28]

Precisely for this reason, Schiaparelli found himself faced with a dilemma. At first he had intended to adhere to Proctor's nomenclature, which Flammarion had already adopted for his 1876 map, with a few changes; for example, Flammarion had decided to retain the older name Mer du Sablier, French for "Hourglass Sea," instead of using Proctor's Kaiser Sea. On looking through his 8.6-inch Merz, however, Schiaparelli found that drastic changes were necessary; some names had to be abandoned, and many new ones had to be introduced to describe the numerous features being seen for the first time. Proctor's four main "continents" were actually a multitude of islands, several of his "seas" had disappeared or shrunk to insignificance (Main Sea, Dawes Sea), while still others had opened up. "In order to avoid misunderstandings and mistakes," Schiaparelli wrote, "I had to create a special nomenclature, which served my particular purpose. This nomenclature, which was devised while I was laboring at the telescope and is probably not without many shortcomings, was retained in my memoir only because it described perfectly what had been seen." [29]

He named the bright and dark areas on Mars after terrestrial lands and seas, and did so without apology, since, he explained,

in general the configurations seen represented such a clear analogy to those of the terrestrial map that it is doubtful whether any other class of names would have been preferable. Do not brevity and clarity also induce us to use such words as *island, isthmus, strait, channel, peninsula, cape,* etc.? Each of which provides a name and description which expresses well what could not otherwise be expressed except through long paraphrases that would need to be repeated each time one spoke of the corresponding object. . . . In order to avoid prejudice regarding the nature of the features on the planet, these names may be regarded as a mere artifice. . . . After all, we speak in a similar way of the seas of the Moon, knowing very well that they do not consist of liquid masses.[30]

Rather than follow Proctor and use the names of past, and in some cases still living, observers of the planet, Schiaparelli drew his appellations from his intimate knowledge of classical literature and the Bible. The ancient Greek founder of physical geography, Dicaearchus, had drawn a line through the middle of his map of the Mediterranean world running from the pillars of Hercules in the west to the Taurus Mountains in the east, which he had called the "great diaphragm." Schiaparelli drew a similar line on Mars running between the belt of dark markings to the south and the lighter regions to the north. The main dark areas, which were given the names of bodies of water, were, proceeding eastward from the Herculis Columnae (Columns of Hercules) in the extreme west: Mare Sirenum (Sea of Sirens), Mare Cimmerium (Sea of the Cimmerians), Mare Tyrrhenum (Tyrrhenian Sea), Mare Hadriaticum (Adriatic Sea), Syrtis Major (Gulf of Sidra), Sinus Sabaeus, Margaritifer Sinus (Pearl-bearing Gulf, the old name for the rich coast of India), Aurorae Sinus (Bay of the Dawn), and Solis Lacus (Lake of the Sun, recalling the legend according to which the Sun rises "in the baths of the ocean"). The bright areas were named for lands. Thus Ausonia (Italy) was separated from Libya by the Tyrrhenian Sea, and other lands included Hellas (Greece), Aeria, Arabia, Eden, Chryse, Tharsis, and Elysium, names which have since become a rich part of Martian lore.

"I do not ask that the [nomenclature] be approved by astronomers in general, nor do I request the honor of its universal acceptance," Schiaparelli wrote modestly. "To the contrary, I am ready to accept as final whichever one is recognized by competent authority. Until then, however, grant me the chimera of these euphonic names, whose sounds awaken in the mind so many beautiful memories."[31] The plain fact of the matter is that Schiaparelli had effectively refashioned Mars with a set of romantic and wistfully evocative names, whose power, despite his stated cautions, was not to be lost on the human capacity to yearn after lost paradises and

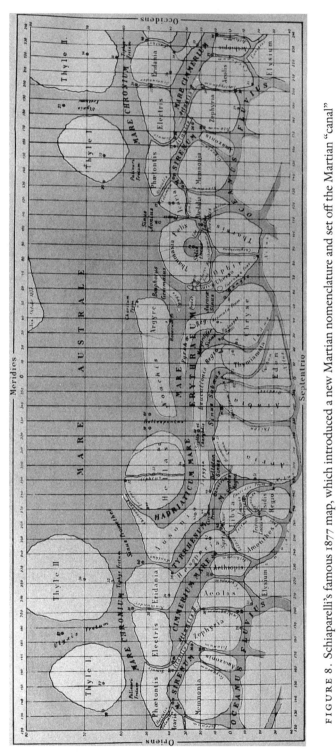

FIGURE 8. Schiaparelli's famous 1877 map, which introduced a new Martian nomenclature and set off the Martian "canal" furor. (From Schiaparelli, "Osservazioni astronomiche e fisiche sull'asse di rotazione e sulla topographia del pianeta Marte," *Atti della R. Accademia dei Lincei*, Memoria 1, ser. 3, vol. 2 [1877–78])

conjure up nostalgic visions. What Percival Lowell once said about naming is no doubt very true: "Naming a thing is man's nearest approach to creating it."[32] In a sense, Schiaparelli's names created a new Mars, or at least a new way of looking at the old Mars. Although it encountered some initial resistance, his nomenclature eventually prevailed. With the single exception of the name Hourglass Sea (Mer du Sablier), which Flammarion continued to prefer to Schiaparelli's Syrtis Major, the French astronomer admitted that Schiaparelli's nomenclature was "euphonic and charming," and he added: "Personally, I hope with all my heart that this ingenious areographical nomenclature will replace all preceding systems."[33] So it did.

It is interesting, and no doubt psychologically significant, that after the introduction of this new map of Mars bearing names so apt to appeal to human emotions, the planet began to gather around itself in succeeding years a considerable mythology of its own. Moreover, it is fitting that this map, whose nomenclature put Mars as much in the realm of mythical as of factual places, should also have been the first to include the strange *canali,* or "canals," which played such an important role in the planet's subsequent mythification (fig. 8).

Indeed, as Schiaparelli continued his observations of the main markings of the planet, smaller details flashed out at him from time to time. There were, for instance, two or three occasions in October 1877 when he witnessed "moments of absolute atmospheric calm. In these circumstances it seemed as if a veil were removed from the surface of the planet, which appeared like a complex embroidery of many colors. But such was the minuteness of these details, and so short the duration of their visibility, that it was not possible to form a stable and sure impression of the thin lines and minute spots therein revealed."[34]

To Schiaparelli's eye these fine details were predominantly linear features, for which he adopted the convenient term that Secchi had first introduced—*canali.* In Italian, *canali* can mean either "channels" or "canals." It is clear that Schiaparelli had completely natural features in mind—indeed, he often used the word *fiume* (river) as a synonym. Strictly speaking, the term *channel* would have been preferable, but instead it was *canal,* with all its connotations of artificial waterways, that was adopted in English, with far-reaching consequences.[35]

Schiaparelli's earlier training in draftsmanship had given him the ability "to transcribe quickly onto paper the almost cinematic impressions of the figures observed in the field of the telescope."[36] However, his eye was "strongly affected by daltonism," or color-blindness; thus, as he himself admitted, he "failed to distinguish gradations of red and green," and he once described the general appearance of the major markings as "almost like that of a chiaroscuro made with Chinese ink upon a general bright background."[37] On the other hand, his color-blindness seems to have made him more sensitive to delicate markings at the threshold of visibility; as a record of fleeting impressions, his observations are unrivaled.[38]

THE CANALS appeared to Schiaparelli only one or two at a time rather than as a whole network. Moreover, and strangely, they were not always best seen when Mars was nearest the Earth but in some cases long afterward; in the words of Percival Lowell, "distance . . . is not, with the canals, the great obliterator."[39] Some of Schiaparelli's own notes in this regard are well worth quoting. On October 4, 1877, when the planet's disk was 21″ of arc across, he recorded in the yellow region between Margaritifer Sinus and Aurorae Sinus only the broad Ganges canal, even though he enjoyed moments of perfect definition. The same area remained unchanged when he studied it again in early November, but on February 24, 1878, on a disk of only 5.7″ arc, he found in this hitherto blank region the Indus, which was now "easily visible."

Though later observers were baffled by these observations, there is a perfectly logical explanation. We now know that dust clouds develop on Mars and are especially frequent during summer in the Martian southern hemisphere, when the planet is near perihelion. Though at times only parts of the surface are covered, at other times the clouds may spread into planet-encircling or even global storms enveloping all the surface features; the latter was the case, for example, in 1971. Schiaparelli made perhaps the first reliable observations of these clouds. Already at the end of September 1877, he had made out a large, bright cloud east of Solis Lacus. On October 10, he found that Mare Erythraeum and Noachis were covered. In beginning his observations that evening, he found that everything appeared normal between 240° and 350° w longitude; he then interrupted his observations to secure a set of measures of a new comet that had been discovered a few days earlier by Wilhelm Tempel at Arcetri. On returning to Mars, with the central meridian now standing at 8° longitude, he wrote in his notebook: "Mars is beautiful. The Mare Erythraeum in large part appears covered by cloud. Noachis is dim. The continent of Deucalion is hardly observable. However, Arabia is plainly in view, and the Sinus Sabaeus stands out as well as ever."[40]

In addition to these clouds, which stood out in contrast to the dark areas beneath them, there seemed to be still others in the bright areas, where "their presence becomes recognizable only in a negative sense—that is, not from what is seen of them, but from what they hide from view."[41] From September to December 1877, the greater part of the planet between the line of the "great diaphragm" and latitude 30° N appeared thus "covered with clouds," including the continental area in which the Indus afterward made its belated appearance. Schiaparelli noted that in the period around opposition, which was shortly before the summer solstice in the southern hemisphere of Mars, clouds and veils were frequent, but that by January, February, and March 1878 the atmosphere of the planet had largely cleared. Thus many of the *canali,* hitherto veiled by mists and clouds, were revealed for the first time despite the much reduced size of the disk.

SCHIAPARELLI WAS NOT the first to see the *canali.* A few ill-defined streaks appear in Schroeter's drawings, and there seems to be at least one in a drawing by Beer and Mädler; still others were recorded by Secchi, Kaiser, and Lockyer, and Dawes was especially prolific in noting them. But with Schiaparelli the *canali* became the dominant motif of the planet, as a simple glance at his map suffices to show. Martian research, long dominated by a simple "analogy to the Earth" approach, and moving confidently forward as observers found further correspondence at each successive stage of discovery, had clearly entered a startling new phase.

Confirmations & Controversies

The next opposition took place in November 1879. Schiaparelli found the winter air at Milan unusually calm and transparent, yielding excellent images. Moreover, he had begun to experiment with new techniques for observation. He illuminated the telescope field to suppress the effects of contrast between the bright planet and the surrounding sky, and by keeping his eye at the eyepiece no longer than necessary to obtain the best views, he was able to avoid eye fatigue and thus work effectively for several consecutive hours when the atmospheric conditions were very good. He also used a yellow filter in front of the eyepiece to improve the contrast of the shadings against the yellowish disk.[1]

Using these precautions Schiaparelli was able to add significant results to the work he had begun at the previous opposition. He obtained micrometric measures of 114 points on the surface—including, on November 10, 1879, a small whitish patch (half a second of arc across) in the Tharsis region. He named it Nix Olympica (the Snows of Olympus) and noted that frequently it seemed to be the site of whitish veils.[2] Also important was his observation that some of the dark areas had changed since 1877. In particular, Syrtis Major seemed to have invaded some of the neighboring bright area of Libya. This seemed consistent with Schiaparelli's "maritime" view of the planet, according to which the dark areas were shallow seas that at times flooded parts of the adjoining lands. As for the canals (as we shall henceforth call them, without apology or quotation marks), they were represented as finer and more regular than they had appeared at first (fig. 9). Moreover, one of them showed up double—the Nilus, between Lunae Lacus and Ceraunius. "To see it as two tracks regular, uniform in appearance, and exactly parallel, came as a great shock," Schiaparelli wrote.[3] This was the first instance of the bizarre process he called "gemination," of which I shall have more to say presently.

Schiaparelli was nothing if not sure of his results, and he wrote of the canals to Nathaniel Green: "It is [as] impossible to doubt their existence as that of the Rhine on the surface of the Earth.[4] Green was also active, observing the 1879 opposition at St. John's Wood, London, though complaining that "the definition afforded by the St. John's Wood atmosphere has barely sufficed to identify the details of the Madeira observations." He was at best lukewarm to the results from Milan, declaring that "faint and diffused tones may be seen in places where Professor Schiaparelli states that new canals appeared during this opposition."[5]

Having embraced (with some changes) Proctor's nomenclature for his own 1877 map, Green was also unenthusiastic about the new Schiaparellian names. "I desire to make an earnest protest against change," he pleaded. "The present names have been in use for many years, they are to be found freely in the publications of the Royal and other astronomical societies, and have been accepted by a large number of workers at the Martial surface. . . . The names Kaiser Sea, De La Rue Ocean, or Dawes' forked bay, are as familiar as household words to those who employ either pen or pencil in this cause, and an affection for old names is at least an excusable weakness."[6]

The striking contrast between the two rival maps of 1877 did not pass unnoticed. Rev. T. W. Webb later pointed out the difference between them:

There is a general want of resemblance that is not easily explained, till, on careful comparison, we find that much may be due to the different mode of viewing the same objects, to the different training of the observers, and to the different principles on which the delineation was undertaken. Green, an accomplished master of form and color, has given a portraiture, the resemblance of which as a whole, commends itself to every eye familiar with the original. The Italian professor, on the other hand, inconvenienced by colour-blindness, but of micrometric vision,

FIGURE 9. Pages from Schiaparelli's observing notebook, 1879.
(Courtesy Luigi Prestinenza)

commenced by actual measurement of sixty-two fundamental points, and carrying on his work with most commendable pertinacity, has plotted a sharply-outlined chart, which, whatever may be its fidelity, no one would at first imagine to be intended as a representation of Mars. His style is as unpleasantly conventional as that of Green indicates the pencil of the artist; the one has produced a picture, the other a plan.[7]

Green had allowed that some of the canals might be the boundaries of faint tones of shade, but he protested that Schiaparelli had not drawn accurately what he had seen. Instead, he had "turned soft and indefinite pieces of shading into clear, sharp lines."[8] Ironically, Schiaparelli himself, who admitted that his 1877 map was "purely schematic," tried in his 1879 map to adopt a different style, "to better approach the true forms one actually sees on the planet, by representing with lines those features which have that appearance in the telescope and with different shades those which exhibit delicate gradations of tint." The result, he hoped, would be "more pleasing to the eye, and more agreeable to the descriptions, than the scheme of pure lines provided with the preceding memoir."[9] This attempt availed little, however, for his representations of the planet in 1882, 1884, 1886, 1888, and 1890 went from strange to bizarre.

At the opposition of 1881, when the apparent diameter of Mars attained only 16″ of arc (compared with 25″ in 1877 and 20″ in 1879), Schiaparelli obtained excellent views of areas of the northern hemisphere that had been presented obliquely at the previous oppositions. Thus, he wrote, "the vast expanses called Oceanus and the Sinus Alcinous, which in 1879 had appeared diffuse and indefinite, seemed to belong to the areas called seas, and were resolved into complex tangles of pure lines."[10] Moreover, the geminations were now widespread; though Schiaparelli recorded only a few in December, they had become alarmingly profuse a month later:

Great was my astonishment on January 19, when, on examining the Jamuna . . . I saw instead of its usual appearance two straight and equal parallel lines running between the Niliacus Lacus and Aurorae Sinus. At first I believed this to be the deception of a tired eye, or perhaps the effect of some kind of strabismus, but I soon convinced myself that the phenomenon was real. From the night of January 19, I passed from surprise to surprise. On the

21st, I discovered the duplication of the Orontes, the Euphrates, the Phison, and the Ganges.[11]

In one month, from January 19 to February 19, 1882, Schiaparelli recorded the gemination of no less than twenty canals. Since most of the doublings had occurred some two months after the Martian vernal equinox, he speculated that they might be a seasonal phenomenon of some sort, although he freely admitted that the explanation "strained the imagination." Nevertheless, whatever their interpretation might be, he was supremely confident that he had seen them. "I have taken all possible precautions to avoid all chance of illusion," he wrote. "I am absolutely certain of what I have observed."[12]

DESPITE THE initial wave of skepticism, the canals, by the mid-1880s, were slowly beginning to gain ground among observers of Mars. Otto von Struve, Wilhelm's son, who had supervised Schiaparelli's training at Pulkova, recorded his own faith without reserve: "I am sorry to say that I have never been able to see the canals, but knowing M. Schiaparelli's excellence as an observer, I cannot doubt that they are there."[13] For many observers the challenge of seeing the canals was irresistible; to fail to do so was to admit observational obtuseness. Under the circumstances, it is hardly surprising that more and more observers saw—or thought they saw—the canals, the whole process bearing a distinct analogy to the story of the emperor's new clothes.

But such an analogy is not quite precise. The Martian "deserts" were not altogether bare. There was undoubtedly *something* in the regions where Schiaparelli drew his network. And even Schiaparelli himself had described a gradual process of recognition:

> In most cases the presence of a canal is first detected in a very vague and indeterminate manner, as a light shading which extends over the surface. This state of affairs is hard to describe exactly, because we are concerned with the limit between visibility and invisibility. Sometimes it seems that the shadings are mere reinforcements of the reddish color which dominates the continents—reinforcements which are at first of low intensity. . . . At other times, the appearance may be more that of a grey, shaded band. . . . It was in one or the other of these indeter-

minate forms that, in 1877, I began to recognize the existence of the Phison (October 4), Ambrosia (September 22), Cyclops (September 15), Enostos (October 20) and many more.[14]

William F. Denning, in 1886, found with his 10-inch (25-cm) reflector that "the more complex and more delicate details of the planet appear, under the most favorable conditions, as *linear shadings, which are extremely feeble, with evident gradations in tone* and with irregularities which produce breaks or condensations."[15] Richard Proctor wrote that "it would be useful if the appearances shown by Schiaparelli could be seen and drawn by observers with real artistic skill. No one who has ever seen Mars through a good telescope will accept the hard and unnatural configurations depicted by Schiaparelli."[16] To which Camille Flammarion added:

The pencil cannot fix features which are scarcely glimpsed. But what else can one do? One can at least distinguish certain shadings, even if it is impossible to be sure of their outlines; and extra detail appears only in rare and fugitive moments of perfect transparency. Illusion or reality? It seems that such telescopic views depend only upon thought. We indicate the features with the pencil, and those which are fugitive, uncertain and perhaps atmospheric take on the same significance as those which are incontestable and permanent.[17]

Among the most widely publicized sightings of the canals were those made by the French observers Henri Perrotin and Louis Thollon, who used the 15-inch (38-cm) refractor at Nice in 1886. Mars was not then in an advantageous position, since even at opposition, on March 6, its apparent diameter never exceeded 14" of arc. Despite their best efforts, the two Frenchmen were long frustrated, but their persistence eventually paid off. Perrotin wrote:

Our first attempts to see the canals were not encouraging, and after several days of fruitless searching, explicable partly by the bad quality of the images and partly because of the actual difficulty of an investigation of this kind, and after once having given up and subsequently returned to the investigation, we were about to abandon the attempt indefinitely when, on April 15, I managed to distinguish one of the canals. . . . Thollon saw it similarly soon afterward. By the end of that night, under good conditions, we had been able to recognize successively several canals presenting, in nearly all respects, almost the character attributed to them by the Director of the Milan Observatory.[18]

Those who seek often do find. François Terby, at his private observatory at Louvain, Belgium, did as so many others did and took the Milan astronomer's latest map along with him to the telescope in order to facilitate identification of the canals: "We took inspiration from the principle announced by some great observers: 'Often,' they say, 'one can see well what one especially seeks.' . . . We have sought the canals of Mars in the regions where we knew that M. Schiaparelli had proved [sic] them to exist . . . and, map in hand, we have patiently and obstinately pursued these very difficult details. It is to this method . . . we owe our partial success."[19]

At this point I cannot resist mentioning some comments by Sir Ernst Gombrich, the noted art critic, about his rather similar experience of straining to interpret garbled radio transmissions during World War II:

I was employed for six years by the British Broadcasting Corporation in their "Monitoring Service," or listening post, where we kept constant watch on radio transmissions from friend and foe. It was in this context that the importance of guided projection in our understanding of symbolic material was brought home to me. Some of the transmissions which interested us most were barely audible, and it became quite an art, or even a sport, to interpret the few whiffs of speech sound that were all we really had on the wax cylinders on which these broadcasts had been recorded. It was then we learned to what extent our knowledge and expectations influence our hearing. You had to know what might be said in order to hear what was said. More exactly, you tried from your knowledge of possibilities certain word combinations and tried projecting them into noises heard. The problem was a twofold one—to think of possibilities and to retain one's critical faculty. . . . For this was the most striking experience of all: once your expectation was firmly set and your conviction settled, you ceased to be aware of your own activity, the noises appeared to fall into place and be transformed into the expected words. So strong was this effect of suggestion that we made it a practice never to tell a colleague our own interpretation if we wanted him to test it. Expectation created illusion.[20]

So it was with Mars. Schiaparelli had taught observers how to see the planet, and eventually it was impossible to see it any other way. Expectation created illusion.

PERROTIN SKETCHED more canals in 1888, now having the advantage of a new telescope, the 30-inch (76-cm) refractor of the Nice Observatory. He also reported that since 1886, Libya, an area about equal in size to France, had been completely inundated by the neighboring sea.[21] Schiaparelli, as we have seen, had reported a partial inundation of the region between 1877 and 1879, and now commented enthusiastically on Perrotin's results: "The planet is not a desert of arid rocks. It *lives;* the development of its life is revealed by a whole system of very complicated transformations, of which some cover areas extreme enough to be visible to the inhabitants of the Earth."[22]

Nevertheless, Perrotin's observations were soon challenged by astronomers using another new telescope even more powerful than the one at Nice—the recently unveiled 36-inch (91-cm) Clark refractor of the Lick Observatory on Mount Hamilton (elevation 4,200 ft, or 1,280 m) in the Coast Range of California.

Though the telescope had been used in some preliminary scouting as early as January 1888, when James Keeler had discovered a new and narrow division in the outer ring of Saturn,[23] the Lick Observatory's official opening was delayed until June 1888. Systematic observations of Mars were not made until July, when the planet was some three months past an only average opposition and the apparent diameter of its disk was only 9″ of arc. Moreover, the planet lay far to the south, never rising more than 30° above the horizon (compared with Saturn, which had been near the zenith), and so was unable to tolerate high magnification. Under the circumstances, nothing sensational could be expected, although the director of the observatory, Asaph Hall's former colleague Edward S. Holden, announced that he, Keeler, and John M. Schaeberle had glimpsed a few of the canals. Holden also declared that "the submerged 'continent' had reappeared . . . and was seen by us here essentially as it has always appeared since 1877. It was most unfortunate that the Lick telescope could not be used for this purpose until so late a date; but it has shown its great power in such work . . . and has conclusively proved that whatever may have been the condition of the 'continent' previous to July it was certainly in its normal condition from that time onward."[24]

This was far from the last word. Despite the great power of the Lick astronomers' telescope, their observations proved to be no less subject than others to the always distracting "personal equation." The representations of Holden and Keeler differed markedly not only from those of Schiaparelli and Perrotin but also from one another, leading Flammarion to write in near despair: "Can one really suppose that it is even the same face of the planet that is being depicted here? . . . What different aspects!"[25]

SCHIAPARELLI USED a new telescope in 1886 and 1888—the 19-inch (49-cm) Merz-Repsold refractor. There is an interesting story concerning this telescope. Following his 1877 observations of Mars, Schiaparelli had been invited to give a lecture to the Academy of the Lynx-eyed in Rome. The talk was well attended, and he was asked to present the lecture again a few days later to the king and queen of Italy at the Quirinal Palace. This time he hinted that with a telescope as large as the great refractor at the U.S. Naval Observatory in Washington, D.C., then the largest in the world, he would very likely be able to find out even more about Mars, "a world little different from our own." He described his address as a "very exciting phantasmagory," and later told Otto Struve that "by employing a little the Flammarionesque style, I managed the affair rather well" (the reference being, of course, to the French astronomer who was well known for his passionate advocacy of the idea of extraterrestrial life).[26] The king and queen were impressed, and when Schiaparelli's request for a larger telescope came before the Chamber of Deputies, it was overwhelmingly approved.

Optically, the 19-inch refractor was not quite of the same high standard as the 8.6-inch (22-cm) Merz, and it suffered from a considerable blue spectrum; nevertheless, Schiaparelli used it almost exclusively in all his observations from May 1886 onward. In 1886 he had failed to see any geminations, but in 1888 they reappeared once again, and the results he got with the large refractor surpassed his expectations. "I believe that I saw the planet well enough on 9, 25 and 27 May," he told Terby,

and I began to be almost satisfied, having confirmed at least three or four geminations. But I had a happy surprise on 2 and 4 June; and only then did I have any idea of the power of a 19-inch aperture for Mars! I then saw that the memorable days of 1879–80 and 1882 had come back for the first time, and that I could again see those prodigious images presented in the telescope field as an engraving on steel; again there was all the magic of the details, and my only regret was to have the disk reduced to 12″ in diameter. Not only could I confirm the gemination of the Nepenthes (*quantum mutatus ab illa!*) and the reappearance of the Triton of 1877, but I could again see Lacus Moeris, reduced to a very small point, but sometimes perfectly visible and scarcely separated from the Syrtis Major.[27]

His views of the Boreo-syrtis and neighboring regions were not as clear, however: "What strange confusion! What can all this mean? Evidently the planet has some fixed geographical details, similar to those of the Earth. . . . Comes a certain moment, all this disappears to be replaced by grotesque polygonations and geminations which, evidently, seem to attach themselves to represent apparently the previous state, but it is a gross mask, and I say almost ridiculous."[28]

IN 1890, Mars was considerably nearer to the Earth than it had been in 1888, but it was much farther south, making observations from Northern Hemisphere observatories difficult. Schiaparelli wrote to Terby that he had discovered a new set of canals around Solis Lacus and that Solis Lacus itself had been "unable to escape the principle of doubling which tyrannizes the entire planet: it is cut crosswise by a yellow band dividing into two unequal parts."[29]

At the Lick Observatory, the results with the largest telescope in the world were no more satisfactory than they had been in 1888. Holden blamed the unusually severe winter weather, which had lasted late into the spring, so that the fine seeing usually found on Mount Hamilton during the summer did not commence until late July or August, by which time Mars was receding from the Earth and too low in the west to be well observed. Nevertheless, he took evident satisfaction in announcing that "the positions of most of Professor Schiaparelli's canals have been verified by some one of us." The endorsement was not quite as resounding as it sounded, however. Only Schaeberle had seen the canals as Schiaparelli did—as narrow lines a second of arc or so in width and in a few cases, at least, double. Keeler and Holden saw only "dark, broad, somewhat diffused bands"; and Holden wondered, as Flammarion had in 1888, "why two observers should agree in their own observations, and should disagree with a third and with the discoverer of the phenomena."[30]

The most important observations of the season were of bright projections into the darkness beyond the terminator line. A visitor on one of the observatory's public nights had first called the attention of Holden, Schaeberle, and Keeler to one of these projections on July 5. Keeler's drawing the next night showed two projections (fig. 10), which presented "much the same appearance as the summits of lunar mountains and craters when first visible outside the terminator of the moon."[31] This plausible explanation was ignored by the press, which made something far more sensational out of them: they were nothing less than signal flashes from the Martians! Inspired by such reports, someone who was looking—perhaps too obsessively—at a canal-filled map of the planet later thought he was able to discern the Hebrew letters making up the word *Shajdai,* the Hebrew name for the Almighty. "This observer was not a devout believer. He was a frank agnostic, and his observation was, therefore, unbiased by any religious zeal," reported the *San Francisco Chronicle:*

> There is a wide field for thought and speculation in this appearance of the name of God standing out unmistakably on the surface of a sister planet. A study of the accompanying illustrations will make it plain how clearly the word stands out. There can be no doubt of the observer's accuracy. The first letter (sheen) is not as sharply defined as are the two others, but washings by the [Martian] ocean have undoubtedly taken place, as is proved by a glance at the original maps, where partly submerged portions of the orange-hued land are indicated. True, the magnitude of the work of cutting the canals into the shape of the name of God is at first thought appalling, but there are terrestrial works which

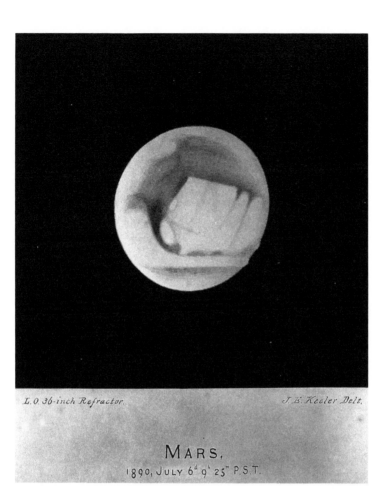

FIGURE 10. A drawing of Mars made by James E. Keeler with the 36-inch refractor of the Lick Observatory on July 6, 1890. Keeler believed the bright projections he saw on this occasion presented "an appearance analogous to that which we observe on the terminator of the Moon"; that is, mountains and craters. Since there are no major relief features in this part of Mars (Cydonia), however, he must have been seeing dust clouds projecting high into the Martian atmosphere. (Courtesy Mary Lea Shane Archives of the Lick Observatory)

to us to-day seem no less impossible. Besides, it is known that the difference in gravitation between Mars and the earth would make it easily possible to do far more work with far less energy on Mars than on the earth.[32]

With reports of Martian signal lights and strange configurations in the Martian deserts all the rage, a French widow, Clara Goguet Guzman, bequeathed 100,000 francs for a prize, named in honor of her son, Pierre Guzman, to be awarded to "the person of whatever nation who will find the means within the next ten years of communicating with a star (planet or otherwise) and of receiving a response."[33] Mars, needless to say, seemed to be the most promising celestial object with which to communicate, and farfetched schemes were promoted by the score over the next few years as practical means of interplanetary communication.[34] One idea was to signal Mars using large letters in the Sahara Desert, inspiring the American astronomer Edward Emerson Barnard to write a fictional account in which Earthlings finally succeeded in sending the Martians the message "Why do you send us signals?" only to receive back the reply, "We do not speak to you at all, we are signaling Saturn."[35]

THE NEXT opposition, in August 1892, was among the most eagerly awaited in the history of Martian studies. At the height of the Mars furor, the planet approached within 35 million miles (56 million km) for the first time since 1877. In that memorable month, Camille Flammarion put the finishing touches to the first volume of his classic *La Planète Mars,* an ambitious compilation and commentary on nearly every study of the planet that had been made up to that time, which remains a mine of information to this day. Flammarion's name has already appeared frequently in these pages; by 1877 he had long been one of the leading students of the red planet (fig. 11).

Flammarion was born at Montigny-le-Roi in the department of Haute Marne in 1842. His interest in astronomy was aroused at the early age of five, when he witnessed a total eclipse of the Sun. At first he seemed destined for the priesthood, but by the time he entered the ecclesiastical seminary of the Cathedral of Langres at age eleven, his passion for astronomical observation was already well developed. A pair of opera glasses had shown him "mountains

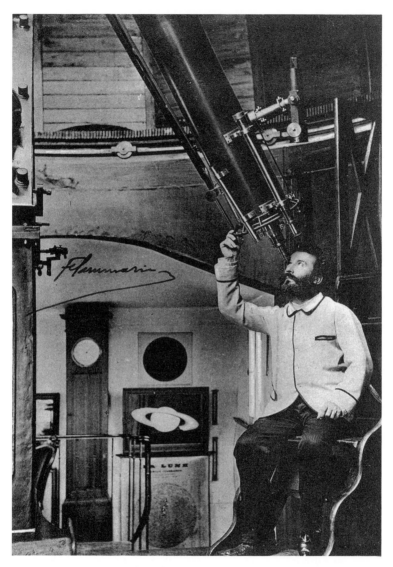

FIGURE 11. Camille Flammarion, observing with the 9-inch Bardou refractor at his Juvisy Observatory. (Courtesy Mary Lea Shane Archives of the Lick Observatory)

in the moon, as on the earth! And seas! And countries! Perchance also inhabitants!"—as he later put it in his rhapsodic style.[36] Henceforth it was clear that if he consecrated himself to any priesthood, it would be a scientific one.

At fourteen he moved to Paris, and there he began writing a book on the origin of the world: *Cosmogonie universelle*. The manuscript was not published at the time, but it came to the attention of U. J. J. Leverrier, the brilliant but irascible director of the Paris Observatory. Leverrier hired Flammarion to work at the observatory as a computer, a job that was not particularly congenial to the romantic young man. At nineteen, Flammarion wrote another book, *La Pluralité des mondes habités*, in which he passionately argued for the existence of extraterrestrial life; it was published in 1862 and immediately became a sensation. This displeased Leverrier, who summoned Flammarion into his office to dismiss him. "I see, Monsieur," he said sternly, "you do not have to remain here. No, it is very simple. You can retire." Flammarion went from the Paris Observatory to the Bureau des Longitudes, but in 1873 Leverrier recalled him, gave him charge of one of the telescopes, and set him to work measuring double stars. In the meantime, however, Flammarion had continued his writing, and his books appeared in quick succession: *Les mondes imaginaires et les mondes réels* and *Marveilles célestes* in 1865, *Études et lectures sur l'astronomie* in 1867, *Voyages en Ballon* in 1868, *L'atmosphere* in 1872, *Lumen* in 1873, and *Le terres du ciel* in 1877. His most successful work, *Astronomie populaire*, was published by his brother Ernest in 1879, and eventually sold 130,000 copies.

In 1882, one of Flammarion's many admirers, a Monsieur Méret of Bordeaux, offered him his chateau—complete with stables, servants' quarters, and parklike setting—at Juvisy-sur-Orge, located about 30 kilometers from Paris by rail. In earlier days this structure had served as a resting place for the kings of France on their journeys between Paris and Fountainbleau, and in one of its rooms Napoleon Bonaparte had learned, on March 30, 1814, of the capitulation of Paris and the downfall of his empire. With his own funds Flammarion proceeded to build an observatory—dedicated, like a temple, to the planet Mars—in which he hoped to realize his dream of discovering extraterrestrial life. He erected a 9-inch (24-cm) Bardou refractor in a dome on the roof of the chateau, and spent May to November of each year at Juvisy; the rest of the year he and his wife, Sylvie, continued to live in their fifth-floor apartment in the Rue Cassini, near the Paris Observatory. In 1887, Flammarion founded the Société Astronomique de France, and served as editor of its monthly publication, *L'Astronomie.*

Sylvie Flammarion furnished some personal details about her remarkable husband to a *San Francisco Call* reporter named Robert Sherard, who visited Paris in May 1893:

> Flam is an extremely methodical man. He gets up regularly every morning at 7 o'clock and spends quite a long time over his toilet. Savants as a rule are a very untidy set, and Flam is an exception to the rule. . . . At a quarter to 8 every morning he has his breakfast, with which he always takes two eggs. From 8 to 12 he works. At noon he has his *déjeuner,* over which he spends a long time. He is a very slow eater. From 1 to 2 he receives visitors, and as he is constantly being consulted on all sorts of questions by Parisian reporters, he is usually kept very busy during this hour. From 2 to 3 he dictates letters to me. . . . At 3 o'clock he goes out and attends to his business as editor of the monthly magazine which he founded and to his duties as a member of various societies. He is back home again at 7:30, when he has dinner and spends the rest of the day in reading. He is a great reader, and tries to keep himself au courant with all that is said on the important topics of the day. At 10 o'clock he goes to bed, for he is a great sleeper.[37]

Though a skillful observer in his own right, Flammarion was obviously far too busy to devote much time to direct telescopic work, and he turned over most of the systematic research at Juvisy to his assistants. In 1892 his staff included Messieurs Guiot, Quénnisset, Schmoll, and Mabir; the next year he was joined by a particularly notable figure, the young E. M. Antoniadi.

Flammarion, in *La Planète Mars,* accepted the maritime view of Mars—that the dark areas were seas and the light areas continents. The existence of seas was shown, he said, by the frequent changes in tone in the dark patches, because such widespread and rapid changes indicated a liquid element rather than solid ground. Less of the surface was covered with water than on Earth, and the Martian seas were clearly "of Mediterranean shallowness." Sometimes they flooded the intermediate regions; at other times they retreated, leaving bare islands. As for the reddish color of the continents, "If the color is of visible ground—that is to say, of the surface," he speculated, ". . . we can assume that the surfaces are sterile and sandy. [But] to us, it seems impossible to condemn a world to a fate of this kind, above all a world in which all the elements of life seem to come together as they do on Mars."[38] Instead, he thought, the reddish color must be from vegetation:

> Why, we may ask, is not the Martian vegetation green? Why should it be?—is the reply. From this point of view, there is no reason to regard the Earth as typical in the universe. Moreover, the terrestrial vegetation can itself be reddish, and has been for the majority of the continents; the first terrestrial plants were lycopods, whose color is a "Martian" reddish yellow. The green substance which gives our vegetation its color—chlorophyll—is made up of two elements; one green, the other yellow. These two elements can be separated by chemical processes. It is therefore perfectly scientific to admit that under conditions different from those on Earth, the yellow chlorophyll can exist alone, or be dominant.[39]

Flammarion maintained that Mars was at a later stage of evolution than Earth, which explained the relative scarcity of water on its surface and the comparative smoothness of its continents.[40] Mountains, he said, are rare, though the terminator projections proved that some peaks exist.

At last he came to the problem of the canals. This, he acknowledged, was "the most delicate part" of his book.[41] Though personally he had succeeded in seeing only the broadest of the canals (Nilosyrtis, Ganges, and Indus), he accepted the existence of the Schiaparellian network. But explanation was not so easy. There was nothing analogous to the canals on Earth. He rejected the idea of the physicist A. Fizeau that they might be open crevasses in immense ice fields, and also the view of E. Penard that they were cracks caused by the cooling of the planet; they were simply too regular to allow such explanations. "On a globe," he asked, "could Nature trace such straight lines, cutting each other in such a fashion? . . . The more we look at these drawings, the less that we can attribute them to blind chance."[42] In the end, he concluded that the canals were watercourses, adding: "The actual conditions on Mars are such that it would be wrong to deny that it could be inhabited by human species whose intelligence and methods of action could be far superior to our own. Neither can we deny that they could have straightened the original rivers and built up a system of canals with the idea of producing a planet-wide circulation system."[43] Such were Flammarion's views in 1892.

FLAMMARION WAS ABLE to make out only a few of the canals at Juvisy in 1892, and another essentially negative report was given by Charles Augustus Young, a distinguished American solar astronomer, who used the 23-inch (58-cm) and 9-inch (23-cm) refractors at the Halsted Observatory in Princeton. Young was a skeptic when it came to the detailed reports of canals given by users of smaller telescopes, and he wrote pointedly: "When I have failed to see with the large instrument anything I supposed I saw with the smaller, it has turned out on examination that the larger instrument was right, and that imagination had constructed a story that was not true by building up faintly visible details and hazy suggestions furnished by the smaller lens."[44]

Perhaps the most famous astronomer to fail to make out any of the canals was Asaph Hall, although in his case the failure was widely and plausibly attributed to the poor atmospheric conditions that prevailed in Foggy Bottom, the unfortunate location of the U.S. Naval Observatory (the telescope was later, in 1893, moved to northwestern Washington, D.C.). Indeed, recall that R. S. Newall, lamenting the dismal results achieved with his large refractor, had commented, "Atmosphere has an immense deal to do with definition."[45] This was especially true when large telescopes were concerned. As Schiaparelli had once pointed out, a star or planet that appeared well defined and quiet in a small telescope with a magnification of 50× became a distorted mass in continual turmoil in a large telescope with a magnification of 500×. Ultimately, Schiaparelli believed, the only solution would be "to put large telescopes on top of tall isolated mountains, such as Teneriffe or Etna, where much of the atmosphere lies under the feet of the observer, and the effects of its agitation can be partly eliminated."[46]

The first actual experiments in this direction were made by Charles Piazzi Smyth, who set up a temporary mountain observatory on the peak of Teneriffe (12,192 ft, or 3,720 m) in 1856. His reports, historian of astronomy Agnes M. Clerke later wrote, "gave countenance to the most sanguine hopes of deliverance, at suitably elevated stations, from some of the oppressive conditions of low-level star-gazing."[47] The first permanent mountaintop observatory was the Lick, established in 1888, followed in 1891 by Harvard Observatory's Arequipa Station, placed at an altitude of 8,100 feet (2,470 m) in the Peruvian Andes. Arequipa was at almost twice the elevation of the Lick Observatory, and its director, William Henry Pickering, claimed that the atmospheric conditions were nearly perfect for planetary observations with the observatory's 13-inch (33-cm) refractor.

Pickering himself quickly became the most prolific—and controversial—observer of the 1892 Mars opposition, which, owing to its far southerly location, was best observed from the Southern Hemisphere. In a year when the Lick observers found it impossible to use more than 350× magnification on their 36-inch refractor because of the planet's low altitude (even so, they succeeded in making out a number of canals), Pickering used 475× and even 1,140×, and reports drawn from his enthusiastic dispatches, printed in the *New York Herald,* were nothing short of sensational:

> September 2. Mars has two mountain ranges near the south pole. Melted snow has collected between them before flowing northward. In the equatorial mountain range, to the north of the gray regions, snow fell on the two summits on August 5 and melted again on August 7.
>
> October 6. Professor Pickering . . . has discovered forty small lakes in Mars.[48]

Pickering published more authoritative reports in the journal *Astronomy and Astro-Physics.*[49] "Many so-called canals exist upon the planet," he wrote, "substantially as drawn by Professor Schiaparelli. Some of them are only a few miles in breadth." He was unable to see even a single gemination, but he did find several "well-developed canals" in the dark areas.[50] This, in his opinion, showed that the latter could not be actual oceans; their greenish tints, he suggested, "might with some show of probability be attributed to the presence . . . of organic life upon the planet."[51] Instead of Flammarion's yellowish red vegetation, Pickering believed in the ordinary greenish variety. Indeed, as we have seen, the idea that the dark areas might be tracts of vegetation had been proposed by E. Liais as early as 1860.

In addition to terminator irregularities such as had been seen at Lick, Pickering made out projections at the limb. In both cases he interpreted the appearances as due to high-altitude clouds rather than mountains. He calculated their heights at around 20 miles (32 km), which made them loftier than the clouds of the Earth—a result, he noted, that was only to be expected given the planet's lesser mass and lower surface gravity.

WITH THE exception of Pickering's results from South America, the opposition of 1892 was a disappointment. It did not lead to developments of significance comparable to those of 1877, when the canals and satellites were discovered. As Clerke summed it up: "The low altitude of the planet practically neutralised [the advantage of distance] for northern observers, and public expectation, which had been raised to the highest pitch by the announcements of sensation-mongers, was somewhat disappointed at the 'meagreness' of the news authentically received from Mars."[52] Yet it was undoubtedly significant that observers using the largest telescope in the world had apparently confirmed Schiaparelli's findings. Clerke, at any rate, was impressed, and concluded that "the 'canals' of Mars are an actually existent and permanent phenomenon."[53]

Lowell

The next opposition of Mars after 1892 occurred on October 20, 1894. At 41 million miles (65 million km), Mars was slightly farther away from the Earth than it had been in 1892. But the greater distance was more than offset by the red planet's greater altitude in northern skies. Indeed, this proved to be one of the most memorable oppositions in the history of Martian exploration—not least because of the emergence on the scene of Percival Lowell, of whom one of his biographers rightly said that "of all the men through history who have posed questions and proposed answers about Mars, [he was] the most influential and by all odds the most controversial."[1] Ideas about Martian life had been floated by Schiaparelli, Flammarion, and others, but it was left to Lowell to fashion them into a coherent whole.

PERCIVAL LOWELL was born in Boston on March 13, 1855, the eldest son of Augustus and Katharine Bigelow Lowell (fig. 12). There was blue blood on both sides of the family. An ancestor, Percival Lowle, had come to what later became known as Newbury, Massachusetts, in 1639. A Bristol merchant, Lowle had followed John Winthrop, the "Lodestone of America," to the Massachusetts Bay Colony at the advanced age of sixty-seven—an undertaking that attests to the almost manic energy that characterized many of his descendants as well.

In addition to their energy, the Lowells tended to exhibit talents for mathematics and literature. Percival Lowell had both in

FIGURE 12. Percival Lowell.
(Courtesy Yerkes Observatory Photographs)

unusual degree, as well as a personal magnetism that many who knew him remarked. Thus journalist Ferris Greenslet wrote: "This reporter has met many of the so-called great men of his time, but none with a more potent personal quality than Percival Lowell. He agrees with another witness that one felt it before, or almost before, he entered the room. It was as if one had been suddenly deposited in a powerful magnetic field."[2]

The early Lowells, although prominent in Massachusetts affairs, were not remarkably wealthy. This changed in 1813 when Francis Cabot Lowell, Percival's great-great-uncle, decided to build a cotton mill at Waltham, Massachusetts, with spinning machinery and a practical power loom modeled on those he had recently seen in the Lancashire textile mills of England. "I well recollect," said Nathan Appleton, a leading stockholder, "the state of admiration and satisfaction with which we sat by the hour watching the beautiful movement of the new and wonderful machine, destined as it evidently was to change the character of all textile industry."[3] In the office of the textile business of his paternal grandfather, John Amory Lowell, Percival Lowell would later work and amass his own fortune. On the other side of the family, his maternal grandfather was Abbott Lawrence, onetime minister to the Court of St. James's. Indeed, the Lawrences were as well-off as the Lowells, with their own textile fortunes. Both families had Massachusetts cities named for them.

Lowell was educated at various private schools in the United States and abroad, and by the age of ten was fluent in French. At eleven he composed a hundred lines of Latin hexameters on the loss of a toy boat. At about this time his father, Augustus, settled the family in the mansion he called Sevenels (so named because there were seven Lowells living there), at 70 Heath Street in Brookline, Massachusetts. Percival no doubt enjoyed rambling about the family estate, which was later affectionately described by his younger sister, Amy, the iconoclast cigar-smoking poet:

Sevenels made a corner where two roads join. . . . I remember them as unfrequented country roads. The place is surrounded by a wall of uncemented pudding-stone over which predatory boys have made it impossible to grow vines. There is an entrance on each road flanked by heavy stone posts, and just inside the wall runs a wide belt of trees, mostly elms, but with just enough evergreens to keep the whole inviolate from the eyes of passers-by.

Within this belt of trees runs a wide meadow, kept from mowing, in June a glory of daisies and buttercups nodding in the wind, a paradise to a child, as I well remember. . . . Beyond the meadow begins the grove. A little handful of land so cunningly cut by paths and with the trees so artfully disposed that one can wander happily among them and almost believe that one is walking in a real wood.

The house itself stands in the midst of lawns and grass terraces. The South Lawn, fringed by trees and bordered with hybrid rhododendrons and azaleas, drops sheerly down to a path, at one end of which is an old fashioned arbour covered with wisteria and trumpet-vines, and two flights of stone steps lead into a formal sunken garden.[4]

It was in the cupola of the roof at Sevenels that fifteen-year-old Percival Lowell set up his first telescope—a 2.25-inch (6-cm) refractor.

Later, at Harvard, Lowell excelled in English composition and mathematics and won the Bowdoin Prize for his essay "The Rank of England as a European Power from the Death of Elizabeth to the Death of Anne." He graduated with honors in 1876, giving, as his part in the commencement exercises, a talk on the nebular hypothesis of Laplace (drawn very heavily from the English philosopher Herbert Spencer). His cousin, the poet James Russell Lowell, referred to Percival as "the most brilliant man in Boston";[5] and one of his professors, the distinguished mathematician Benjamin Peirce, invited him to stay on at Harvard to teach mathematics. Lowell declined—"because I preferred not to tie myself down," he later recalled, "not because mathematics had not appealed to me as the thing most worthy of thought in the world."[6] Instead, with his cousin and freshman roommate, Harcourt Amory, he opted for a Grand Tour of Europe as far as Syria, and though he had so recently refused to tie himself down to "the thing most worthy of thought in the world," he almost managed to get himself enlisted for the front in the war between Serbia and Turkey.

On returning to the United States in 1877, the year of Schiaparelli's fateful discovery of the Martian canals, still having no impulse to a profession, Lowell went to work in his grandfather's textile business, where he remained for six years, learning the ways of business, for a time acting as treasurer—that is, as the executive head of a cotton mill—and managing trusts and electric companies. In 1883 he left the family business and set sail for the Far East, the first of three such voyages he would make over the next decade. The circumstances surrounding what must have seemed a sudden decision were in some ways similar to those that later led him to turn to Mars. Inspired by the 1882 lecture series on Japanese culture given by zoologist Edward Sylvester Morse at the Lowell Institute (founded by Percival's great-uncle, John Lowell, and run by his father, Augustus Lowell), he was seized by a sudden enthusiasm to join Morse's crusade to preserve traditional Japanese culture in the face of rapid modernization.[7] In Tokyo, he rented a house, hired Japanese servants, and began to learn the language.

When he was approached to accompany a special trade mission from Korea to the United States, he hesitated, "mainly due," as his brother Abbott Lawrence Lowell later put it, "to anxiety to what his father would say."[8] However, he received encouragement from another cousin, William Sturgis Bigelow, a Boston physician who shared his enthusiasm for the Orient. Bigelow wrote to Augustus Lowell that although Percival "distrusts himself too much, he has great ability, he has learned Japanese faster than I ever saw any man learn a language—and he only needs to be assured that he is doing the right thing to make a success of anything he undertakes, whether science or diplomacy."[9] (In fact, Lowell's accomplishments as a linguist were somewhat exaggerated, for apparently he never did succeed in achieving either fluency or literacy in Japanese.)[10] Nevertheless, the young Lowell overcame any opposition on his father's part and any doubts of his own abilities that he might have harbored, for, as Ferris Greenslet remarked, "this was the last occasion in which there is any record of such distrust."[11]

After the Korean mission completed its aim of establishing trade relations with the United States, Lowell returned to Seoul with the Korean delegation, and, as a diplomatic official, was given a house forming part of the Foreign Office. The situation in Seoul must have been quite appealing to someone with Lowell's poetic temperament, as his description of his house there indicates: "From the street you enter a courtyard, then another, then a garden, and so on, wall after wall, until you have left the outside world far behind and are in a labyrinth of your own."[12]

Lowell remained in the Far East for ten years, returning home only to see his books through the publication process: *Chosön: The Land of the Morning Calm* in 1886, *The Soul of the Far East* in 1888, and *Noto* in 1891.[13]

Despite his strong attraction to Japanese art and gardens, his romantic impulse for the Far East was soon tempered by his irritation at the inefficiency and irrationality of premodern people. His 1891 trip to the peninsula of Noto, for example, concluded with the following reflection: "From below, by the river's mouth, the roar of the surf came forebodingly up out of the ashen east; but in the west was still a glory, and as I turned to it I seemed to look down the long vista of the journey to western Noto by the sea. I thought how I had pictured it to myself before starting, and then how little the facts had fitted the fancy."[14]

In the last of his Far Eastern books, *Occult Japan* (1895), Lowell took up the subject of Shinto trances, which he had first witnessed on the summit of Mount Ontaké, and described the séances he had held in his own house ("red flame and potent spells in a dark dark room," he described them to his friend Frederic Jepson Stimson). His experiments included rather cold-blooded attempts to test the sensibility of the trance subjects by sticking pins in them. The journalist Lafcadio Hearn, who had greatly admired Lowell's earlier work and who spent most of his own adult life in Japan, characterized *Occult Japan* as "painfully unsympathetic, Mephistophilean in a way that chills me."[15]

As LOWELL'S infatuation with the Far East began to wane, his boyhood interest in astronomy came once more to the foreground. It was premonitory of the change in direction his career was about to take that on his last voyage to Japan, in late 1892, he carried with him a 6-inch (15-cm) Clark refractor, which he used to observe Saturn from his Tokyo residence—Mars being neglected only because it was already past opposition by that time.

Indeed, as early as 1890 he had begun to correspond about Mars with William H. Pickering. After the 1892 opposition, Pickering was dismissed from his post at Arequipa by his older brother, Edward C. Pickering, the director of the Harvard College Observatory, who disapproved of his brother's rather sensationalistic results about Mars (William had been sent to Arequipa primarily to obtain stellar spectra, not to observe planets) and his inability to stay within his budget. On returning to Boston, William tried to interest Edward in sending a Harvard expedition to the Arizona Territory, where he expected to find excellent seeing, in order to observe the coming opposition of Mars in October 1894. When that effort failed, he immediately began casting around for other funding sources. He had not yet succeeded in lining up any firm support when, in January 1894, he heard from Percival Lowell.

Lowell's decision to take up astronomy in earnest, though it had been long in developing, apparently came rather suddenly. As late as October 1893, as he was preparing to leave Tokyo for the last time, he wrote to artist Ralph Curtis that he was considering an Easter jaunt to Seville in the spring of 1894. He returned to Boston in December 1893 and received Camille Flammarion's *La Planète Mars* as a Christmas gift from his aunt, Mary Putnam, read it at lightning speed, and scrawled "Hurry" across the page.[16] The Lowell family motto was *Occasionem cognosce*—"Seize your opportunity." Lowell realized that the scientific investigation of Mars presented such an opportunity, but the last really favorable opposition of the century was fast approaching. He would indeed need to hurry if he was to take advantage of it.

In the throes of his new enthusiasm, Lowell arranged to meet Pickering on or about January 17, 1894. The intermediary for the meeting was probably Lowell's cousin Abbott Lawrence Rotch, an amateur meteorologist who in 1885 had founded the Blue Hill Meteorological Observatory outside Boston and had just returned with Pickering from Peru, where he had set up a meteorological station at Arequipa. Within a week of the meeting, Lowell and Pickering had come to terms on an expedition of the Harvard Observatory to Arizona for the purpose of observing Mars. At first, both of the Pickering brothers seem to have assumed that Lowell intended to act primarily in the role of patron, but Lowell soon made it clear that he had no interest in merely "going along"; he fully intended to keep control of the expedition himself. The agreement with Harvard was somewhat acrimoniously dissolved, and Lowell worked out a different arrangement whereby William Pickering and Andrew Ellicott Douglass, a young graduate of Connecticut's Trinity College who had been Pickering's assistant at Arequipa, would be granted one-year leaves of absence from Harvard so that they could accompany him to Arizona, in return for which Lowell would personally pay their salaries.[17] Seth Carlo Chandler, Jr., wrote to E. S. Holden at the Lick Observatory that Edward Pickering had attempted to capture the new observatory for Harvard, "but found he had caught a Tartar in Lowell who is one who can see through a millstone and not one it is safe to play as a sucker."[18] Henceforth the observatory would be known eponymously as the Lowell Observatory.

With Lowell having established firm control, the expedition to Arizona now moved ahead. For telescopes, they would borrow a 12-inch (30-cm) refractor from Harvard and take along an 18-inch (46-cm) refractor recently finished by optician John Brashear in Pittsburgh. Pickering drew up designs for a lightweight dome to be made of canvas and wood. Meanwhile, the location of the observatory—in many ways the crucial factor—had yet to be chosen. Douglass was sent west in early March to scout out sites, taking along the 6-inch refractor Lowell had used in Japan. With this instrument he planned to test the seeing, using a ten-point scale developed by Pickering at Arequipa that was based on the appearance of a bright star's diffraction disk and rings.[19] Douglass arrived in Tombstone by March 8 and tested the seeing there, then went on to Tucson, Tempe, and Phoenix in southern Arizona before veering north to Prescott and Ash Fork. Finally he came to Flagstaff, on the main Santa Fe Railroad line to California. The altitude at this site (7,000 ft, or 2,100 m) on the Coconino Plateau appealed to Lowell, who wrote Douglass that "other things being equal, the higher we can get the better."[20] Douglass found the results good in the few observations he made in the "opening in the woods" on the mesa just west of town, but only marginally better than those he had obtained elsewhere in Arizona. Though Douglass wanted to carry out additional tests, time was running out. Thus Lowell, on April 16, decided to build the observatory in Flagstaff.

WHAT PERCIVAL LOWELL hoped to accomplish through this "speculative, highly sensational and idiosyncratic project" is well documented in an address he gave to the Boston Scientific Society on May 22, 1894, which was printed in the *Boston Commonwealth*.[21] His main object, he stated, was to study the solar system: "This may be put popularly as an investigation into the condition of life on other worlds, including last but not least their habitability by beings like [or] unlike man. This is not the chimerical search some may suppose. On the contrary, there is strong reason to believe that we are on the eve of pretty definite discovery in the matter." To Lowell, the implications of Schiaparelli's network of *canali* were self-evident:

Speculation has been singularly fruitful as to what these markings on our next to nearest neighbor in space may mean. Each astronomer holds a different pet theory on the subject, and pooh-poohs those of all the others. Nevertheless, the most self-evident explanation from the markings themselves is probably the true one; namely, that in them we are looking upon the result of the work of some sort of intelligent beings. . . . [T]he amazing blue network on Mars hints that one planet besides our own is actually inhabited now.

With this lecture Lowell had "taken the popular side of the most popular scientific question about," as W. W. Campbell would later declare,[22] and had—in rather unscientific fashion—announced his conclusions before he had even put eye to eyepiece.

Lowell documented his arrival at Flagstaff, on May 28, 1894, in a letter to his mother, to whom he wrote nearly every day: "Here on the day. Telescope ready for its Arizonian virgin view. . . . Today has been cloudy but now shows signs of a beautiful night and so, not to bed, but to post and then to gaze."[23] The two telescopes, the 18-inch Brashear and the 12-inch refractor borrowed from Harvard, were mounted together inside the dome designed by Pickering (which still lacked its canvas cover). The astronomers—Lowell, Pickering, and Douglass—had taken temporary accommodations in a hotel in town, but Lowell and Pickering set up camp in the dome that night to be ready for "early rising Mars." However, they were clouded out, and rain fell through the open dome. Finally, on May 31, the sky cooperated, and they used the 12-inch refractor for their first view of the planet. The following night Lowell recorded his first impressions with the 18-inch refractor: "Southern Sea at end first and Hourglass Sea . . . about equally intense. . . . Terminator shaded, limb sharp and mist-covered forked-bay vanishes like river in desert."[24] Lowell's use of the word *desert* is remarkable, for it contains the kernel of his later theory about the planet. Perhaps Lowell's imagination had already been captured by the lonely desert south of Flagstaff, whose appearance as seen from one of the San Francisco Range peaks he later described:

The resemblance of its lambent saffron to the telescopic tints of the Martian globe is strikingly impressive. Far forest and still farther desert are transmuted by distance into mere washes of color, the one robin's-egg blue, the other roseate ochre, and so bathed, both, in the flood of sunshine from out of a cloudless burnished sky that their tints rival those of a fire-opal. None otherwise do the Martian colors stand out upon the disk at the far end of the journey down the telescope's tube. Even in its mottlings the one expanse recalls the other.[25]

Lowell and Pickering recorded the canal Lethes on June 7, and Lowell gloated: "This is the first canal seen here this opposition, and in all likelihood the first seen anywhere." Two days later he found it "very broad and glimpsed double," but still a novice to the astronomical trade, he doubted himself: "These sudden revelation peeps may or may not be the truth." On June 19 he wrote: "With the best will in the world I can certainly see no canals."[26]

After only a month of observing, Lowell returned to Boston, leaving the observatory in the hands of Pickering and Douglass. In Lowell's absence, Pickering attempted to measure the polarization of light from one of the dark areas but found that the reflected light from Mars was not polarized. This meant that the area was probably not covered with water.[27] Douglass, meanwhile, was verifying Pickering's 1892 observation that there were canals crisscrossing the dark areas. The evidence was clearly growing that these areas were not seas after all. But if not seas, then what?

Lowell returned to Flagstaff in early August. When the observations of Mars had begun, in June, it had been early spring in the Martian southern hemisphere; now it was getting into high summer—the soufflé-thin polar cap was melting rapidly, and Lowell noted a dark band around it, which he described as a "ring of antarctic ocean."[28] Later, the antarctic sea also disappeared, while corresponding changes took place in the dark areas located in the southern hemisphere—what had at first appeared as a uniform dark belt stretching "unbroken from the Hourglass Sea [Syrtis Major] to the columns of Hercules" had begun to break up into islands and peninsulas. "It will at once be seen how this bit of evidence fits into the other," Lowell suggested. "If the polar sea were thus to descend in a vast freshet toward the equator such are the appearances the freshet might be expected to present. As the water progressed farther and farther north the regions it left behind would gradually dry up, and from having appeared greenish-blue would take on an arid reddish-yellow tint; precisely what is observed to take place."[29] Regarding whether the changes in tint from blue-green to reddish yellow were due to the arrival and subsequent retreat of water itself or merely its vapor, Lowell hedged. "Either water itself or vegetation its consequence would show us the first color while deserts would show the last," he wrote. "The configuration of the seas would likewise be explicable on either hypothesis. It would merely be a question of present seas or past sea-bottoms. . . . The probability is that these areas are in part water, in part fertile land."[30] Although he had arrived in Flagstaff accepting the Schiaparellian maritime theory of the dark areas on Mars, he was now embracing the Liais-Pickering vegetation theory.

The canals too, which he was now sketching prolifically and in essentially Schiaparellian form, followed the general development of the broader features of the disk. "They do not all begin to develop at the same season," he announced. "Those nearest the south pole start first. The Solis Lacus is the one to lead off the list. Then the others follow in their order north. . . . Some weeks elapse after the water has to all appearance gone down the disk before the canals appear; a delay of just about the length of time it would take vegetation to sprout."[31] The dark spots at the juncture of the canals, which Pickering in 1892 had called lakes, were also seen to change, not in size but in color. They seemed to deepen in hue, and this too was suggestive:

> When we put all these facts together, the presence of the spots at the junctions of the canals, their apparent invariability [in] size, their seasonal darkening, and last but not least the resemblance of the great equatorial regions of Mars to the deserts of our Earth, one solution instantly suggests itself of their character, to wit: that they are oases in the midst of that desert.
>
> Here then we have an end and reason for the existence of canals and the most natural one conceivable—namely that the canals are constructed for the express purpose of fertilizing the oases. . . . And just such inference of design is in keeping with the curiously systematic arrangement of the canals themselves. . . . The whole system is trigonometric to a degree.[32]

In short, Mars was a world well on the way to utter desiccation. It was inhabited, and its inhabitants, in order to survive, had had to built a vast system of irrigation canals to transport precious water from the melting polar caps. This, in a nutshell, was Lowell's "theory," and needless to say, it created an immediate sensation.

Despite having observed the planet through only a single opposition, Lowell, in a spurt of manic energy, blitzed the press with his sensational results. He had begun publishing a series of articles on Mars in W. W. Payne's journal *Popular Astronomy* even before the opposition was over, and he followed that up with a similar series in the *Atlantic Monthly*. In February 1895 he gave a series of well-attended lectures in Boston's Huntington Hall, then concluded, finally, in December 1895 with his first book on the red planet, entitled simply *Mars,* in which he described in detail his observations and conclusions therefrom. Lowell's drawings and maps were even stranger than those Schiaparelli had published. To some extent this can be attributed to the fact that Lowell, for all his undoubted and diverse talents, was, as Carl Sagan has pointed out, "unfortunately one of the worst draftsmen who ever sat down at the telescope and the kind of Mars that he drew was composed of little polygonal blocks connected by a multitude of straight lines."[33] To that I would add only that Pickering and Douglass were, if anything, even more maladept draftsmen!

Lowell's theory of intelligent life on Mars unleashed a firestorm of controversy. The public was fascinated, while professional astronomers generally viewed him with suspicion; some were openly hostile. Lowell had opened his *Popular Astronomy* series by attributing the failure of skeptics to see the evidence to the fact that they had not observed the planet in steady enough air. He added: "No amateur need despair of getting interesting observations because of the relative smallness of his object-glass. . . . In matters of planetary detail size of aperture is not the all-essential thing. . . . A large glass in poor air will not show what a small glass will in good air."[34] Undoubtedly Lowell was a fast study, but it is still remarkable that he could write this after only one month of observing Mars at Flagstaff! Later, he condemned the use of large reflecting telescopes for planetary work as well, calling them "well-nigh worthless."[35] After reading such sweeping comments, James Keeler, then at the Allegheny Observatory in Pittsburgh, complained to George Ellery Hale: "I dislike his style. . . . It is dogmatic and amateurish. One would think he was the first man to use a telescope on Mars, and that he was entitled to decide offhand questions relating to the efficiency of instruments; and he draws no line between what he sees and what he infers." [36] Hale and Keeler, who were the coeditors of the influential *Astrophysical Journal,* eventually declined to publish any of Lowell's submissions in their journal.

Lowell's Martian theory immediately made him one of the most prominent men in Boston, and he did not appear to mind the celebrity, at least judging by a vignette from about this time related by Ferris Greenslet:

> He had bought for his life *en garçon* a small high house on the upper side of West Cedar Street. There during the winter a young editor and publisher from New York, passing with a bag of manuscripts to his own modest establishment in the next block, used to observe him every weekday at five-thirty. His handsome head was to be seen *vis-à-vis* the *Boston Evening Transcript* beneath a life-sized plaster Venus similar to those that infest the Athenaeum. Visibility was perfect, for the shade was always raised to the very top of the window as if to admit no impediment to a message from Mars.[37]

After seeing his book *Mars* through the press in December 1895, Lowell set sail for Europe to confer with eminent Martian observers there. In Paris, he dined with Flammarion in his apartment on the Rue Cassini, describing the occasion in a letter to his father: "There were fourteen of us, and all that could sat in chairs of the zodiac, under a ceiling of pale blue sky, appropriately dotted with fleecy clouds, and indeed most prettily painted. Flammarion is nothing if not astronomical. His whole apartment, which is itself *au cinquieme,* blossoms with such decoration." [38] In addition to their mutual interest in Mars and extraterrestrial life, the two men shared a fascination with the occult—Lowell, as we have seen, had held séances in his house in Japan, and Flammarion sometimes invited mediums to his Paris apartment. Flammarion had from the first commended Lowell for founding "an observatory inspired, as that at Juvisy, with the dominant idea of studying the conditions of life on the surface of the planets of our system." [39] He subsequently recognized the importance of Lowell's Martian observations, which were, he wrote, "of the highest interest, though certainly controversial, and they advance our knowledge of the planet, even if we do not accept them as definitive."[40] He could not, however, bring himself to accept Lowell's conclusion that the dark areas were tracts of vegetation. He still regarded them as seas, although, he admitted, they were in many places probably little more than marshes.[41] From Paris, Lowell traveled on to Milan, where he met the man he admired above all others and always referred to as *cher maître Martien*.

What Schiaparelli's expectations of the American may have been are less certain, though he later confided to François Terby: "It is certain that Lowell commands superior means to any hitherto employed on Mars. If his perseverance and enthusiasm do not desert him, he will make considerable contributions to areography; on the other hand, he needs more experience, and must rein in his imagination."[42]

At about this time the Milan astronomer was becoming increasingly aware of problems with his eyesight; while observing Mars, he confided to Terby, he was troubled by "a diminution of the sensibility to weak illuminations; I attribute this to the observations of Mercury near the Sun carried out from 1882 to 1890."[43] Moreover, the air over the growing and increasingly polluted city of Milan was no longer as tranquil as it once was. Finally, in May 1898, Schiaparelli announced his retirement from observational work:

> The time has come to let others take over the careful study of the phenomena of Mars. I will publish my observations, in the hope that time will resolve the feelings of doubt and distrust with which they are received by nearly one and all. Whoever wishes to study Mars successfully, must have a keen eye (like my left eye; my right is useless for observations), and must work in a calm atmosphere with a telescope that is able to concentrate the rays in the red part of the spectrum—the other rays must be eliminated with a colored glass. Add to this as prerequisites long practice and great prudence in the conclusions one draws from the observations.[44]

But though his eyes were no longer keen enough to place him in the first rank of planetary observers—indeed, he published none of the observations he made after 1890—Schiaparelli nevertheless remained the leading authority on Mars, and his pronouncements on the subject were eagerly awaited. In 1893 he penned a widely quoted paper titled "Il Pianeta Marte," in which he suggested that the canals were in all probability natural features produced during the evolution of the planet—perhaps similar to the English Channel or the Channel of Mozambique. When he turned to a consideration of the geminations, however, he admitted that it was difficult to think of a natural explanation: "Their singular aspect, and their being drawn with absolute geometrical precision, as if they were the work of rule or compass, has led some to see in them the work of intelligent beings, inhabitants of the planet. I am very careful not to combat this supposition, which includes nothing impossible."[45]

By 1895, Schiaparelli was not only "careful not to combat" the supposition, he seems to have embraced it. That year he published another widely quoted paper, "La Vita sul Pianeta Marte" (The life of the planet Mars), in which he wrote that the idea that the geminations were perhaps best explained as owing to the activity of intelligent beings "ought not to be regarded as an absurdity."[46] On the contrary, he said, "one cannot [otherwise] comprehend how in the same valley the moisture and vegetation sometimes make a single line, in other cases two parallel lines of unequal breadth and separated by unequal intervals, between which remains a sterile space deprived of water. Here, the intervention of intelligent thought seems well indicated."[47]

He proceeded to work out the details of a system of locks and dikes that would both regulate the water flow on Mars for the convenience of the inhabitants and also satisfy the observations made from Earth. A remarkable performance! Yet Schiaparelli insisted that he did not mean to be taken seriously and closed with the comment, "I leave now to any lecturer who cares to do so to continue these considerations; as for me, I am descended from a hippogriff."[48] To Flammarion he sent a copy of the paper, on which he wrote at the top of the page, "Semel in anno licet insanire" (Once a year it is permissible to act like a madman).[49] Once again he remained, to the exasperation of his contemporaries, impenetrable when it came to his true views about the nature of the canals.

Though he later gave Lowell's theories a sympathetic hearing, writing in 1897 that the system of canals "presents an indescribable simplicity and symmetry which cannot be the work of chance," and telling Lowell himself, "Your theory of vegetation becomes more and more probable,"[50] he refused to commit himself. To inquiries about the nature of the canals, he continued to respond, "I don't know!"[51] In an 1899 review of Lowell's observations, he struck an all-too-familiar note of agnosticism, writing that the nature of the canals was still "entirely obscure, despite the theories, oftentimes pretty and very ingenious, which they have occasioned."[52] Twenty-two years had passed since he had discovered the remarkable network, yet it remained a source of bewilderment to him. In this respect he was far from being alone. E. M. Antoniadi, Flammarion's assistant at Juvisy, wrote of the "canal deadlock" and recalled this as a time when "everything was darkness to all."[53]

How the Eye Interprets

There were important developments in 1894 other than the founding of the observatory at Flagstaff and the emergence of Percival Lowell as the central figure in the debate over the Martian canals. The spectroscopic detection of water vapor on Mars had been confidently announced in the 1860s and 1870s by William Huggins in England, Jules Janssen in France, and Hermann Vogel in Germany. Janssen's spectroscopic observations in 1867 were especially noteworthy because he had carried his instruments to the summit of 10,741-foot (3,280-m) Mount Etna in Sicily, above much of the lower atmosphere of the Earth, and had noted what seemed to be a definite intensification of the lines caused by water vapor in the spectrum of Mars compared with that of the airless Moon. In 1894 this result was disputed by W. W. Campbell at Lick. With an improved spectroscope attached to the Lick 36-inch (91-cm) refractor in the dry summer air over Mount Hamilton (fig. 13), Campbell found that "the spectrum of Mars . . . appeared to be identical with that of the Moon in every respect."[1] The direct evidence of water vapor on Mars had vanished into thin air.

EVEN MORE damaging to the Lowellian view of the planet were the observations of Campbell's colleague at Mount Hamilton, Edward Emerson Barnard, who was already well known for the keenness of his sight and the soundness of his judgment. Barnard had overcome great hardships in achieving his leading position in astronomy. He had grown up poor in Nashville, Tennessee, after

FIGURE 13. W. W. Campbell with the 36-inch spectroscope he used in the summer of 1894 in his quest for evidence of water vapor in the atmosphere of Mars. (Courtesy Mary Lea Shane Archives of the Lick Observatory)

the Civil War. At the age of nine, after only two months of formal schooling, he had been sent to work in a photograph gallery, where he was assigned the job of keeping a large solar camera used for portrait photography pointed toward the Sun by manually turning a set of large wheels. He was still working at the photograph gallery as a printer and operator when, at the age of nineteen, he received as a surety of a loan a volume of astronomical writings by Rev. Thomas Dick. The book immediately awakened a passion for astronomy in Barnard, and with the help of an older colleague at the photograph gallery, James W. Braid, Barnard put together a small telescope from a simple tube and an old spyglass lens found in the street, which, he later recalled, "looked as if [it] had been chipped out of a tumbler by an Indian in the days of the Mound builders." Nevertheless, this telescope "filled my soul with enthusiasm when I detected the larger lunar mountains and craters, and caught a glimpse of one of the moons of Jupiter."[2] By 1877 he had saved enough money to buy a better telescope, a 5-inch (13-cm) refractor, and he studied Mars at the oppositions of that year and, especially, 1879. In 1880, he wrote a small book about the planet, unfortunately never published, which advises an admirable caution in interpreting the observations of the planet:

> It is well to fetter the wings of our fancy and restrain its flights. It is quite possible we may have formed entirely erroneous ideas of what we actually see. The greenish gray patches may not be seas at all, nor the ruddy continents, solid land. Neither may the obscuring patches be clouds of vapor. Man is too quick at forming conclusions. Let him but indistinctly see a thing, or even be undecided as to whether he does actually see it and he will then and there set himself to theorizing, and build immense castles of conjecture on a foundation, of whose existence he is by no means certain.[3]

After discovering numerous comets and completing a fellowship in astronomy at Vanderbilt University, Barnard was hired by E. S. Holden for the original staff at Lick Observatory. In 1892 he began using the 36-inch refractor on a regular basis (every Friday night). With this telescope he found the notoriously difficult fifth satellite of Jupiter in September 1892; it was the last satellite in the solar system to be discovered visually. He also studied Mars at the 1892 opposition, but the planet's far southerly declination made observing conditions very poor, and he was unable to use magnifications higher than 350× on the 36-inch refractor, or so he told Schiaparelli during a visit to Europe in the summer of 1893.[4]

In 1894, with Mars higher in the sky above Mount Hamilton, Barnard obtained more satisfactory observations of the planet. Already in July 1894—three months before opposition—he had begun to obtain breathtaking views of the surface details. On July 23, for instance, with Mare Sirenum in view, he recorded two tantalizing, small dusky spots, which appeared "very feeble and faint when near the middle of [the] disc" but grew black as they drew near the terminator. By early September he was drawing Mars on a scale of 5 inches (13 cm) to the planet's diameter, and watching it from sunset till dawn. The seeing grew exceptionally steady at times, allowing him to use magnifications of more than 1,000×. By now the Solis Lacus region had come into view, and on September 2–3 he recorded in his notebook: "There is a vast amount of detail. . . . I however have failed to see any of Schiaparelli's canals as straight narrow lines. In the regions of some of the canals near Lacus Solis there are details—some of a streaky nature but they are broad, diffused and irregular and under the best conditions could never be taken for the so called canals."[5] A week later he had another chance at Mars with the great refractor, but because there was no water in the engines he had to turn the dome and wind the clock drive by hand—"dreadfully hard and exhausting work," he noted.[6] The magnificent Hourglass Sea—Schiaparelli's Syrtis Major—was coming onto the disk. The next day Barnard confided to Simon Newcomb:

I have been watching and drawing the surface of Mars. It is wonderfully full of detail. There is certainly no question about there being mountains and large greatly elevated plateaus. To save my soul I can't believe in the canals as Schiaparelli draws them. I see details where some of his canals are, but they are not straight lines at all. When best seen these details are very irregular and broken up—that is, some of the regions of his canals; I verily believe—for all the verifications—that the canals as depicted by Schiaparelli are a fallacy and that they will be so proved before many oppositions are past.[7]

Barnard's first published report of what he had seen on Mars with the great refractor did not appear until two years later, when he made reference to it almost incidentally in a discussion of his 1894–95 Saturn observations. An English amateur named A. Stanley Williams had published reports of faint Saturnian spots observed with only a 6.5-inch (16-cm) reflector, and Barnard, who had seen nothing of the sort, disputed the claims, taking advantage of the opportunity to urge his strong conviction regarding the superiority of large instruments for planetary work. In particular, he declared his skepticism about the canal-filled drawings of Mars made by observers (including Williams himself) using small telescopes, and went on to describe the very different results he had obtained with the great refractor in 1894:

> On several occasions during that summer, principally when the planet was on the meridian shortly after sunrise—at which time the conditions . . . are often exceptionally fine at Mount Hamilton—its surface has shown a wonderful clearness and amount of detail. This detail, however, was so intricate, small, and abundant, that it baffled all attempts to properly delineate it. Though much detail was shown on the bright "continental" regions, the greater amount was visible on the so-called "seas." Under the best conditions these dark regions, which are always shown with smaller telescopes as of nearly uniform shade, broke up into a vast amount of very fine details.[8]

Though he found these details impossible to draw and difficult to describe, Barnard saw, or seemed to see, analogies to the rugged terrain around Mount Hamilton itself:

> To those, however, who have looked down upon a mountainous country from a considerable elevation, perhaps some conception of the appearance presented by these dark regions may be had. From what I know of the appearance of the country about Mount Hamilton as seen from the observatory, I can imagine that, as viewed from a very great elevation, this region, broken by canyon and slope and ridge, would look just like the surface of these Martian "seas."[9]

No one had ever seen Mars so clearly. The views Barnard had after sunrise with the 36-inch refractor revealed a scene utterly unlike the hard, sharp features shown in the average drawings of the day. This new Mars was a revelation. There was nothing artificial looking anywhere on the surface—on this last point, at least, Barnard had no doubt.

THE LAST important development during the memorable year 1894 was the publication of a provocative paper by the English astronomer Edward Walter Maunder (fig. 14). As far back as 1879,

FIGURE 14. Edward Walter Maunder.
(Courtesy Mary Lea Shane Archives of the Lick Observatory)

Maunder had made out, at least in qualified form, some of the Schiaparellian canals, though he had tended to agree with Nathaniel Green's interpretation of them. "Where I have represented shaded districts, [Schiaparelli] has drawn hard lines corresponding with the borders of those districts," Maunder had declared in 1882, "so that where he has given a number of parallel and interlacing lines, I should myself have rather shown faint shaded districts between those lines."[10] He now looked at the canal question from a new perspective that went far toward resolving the "canal deadlock."

Born in 1851, the youngest son of a Wesleyan minister, Maunder attended King's College, London, and after graduating worked for a time in a London bank. His real chance came in 1873, when the position of photographic and spectroscopic assistant at the Royal Observatory at Greenwich became vacant. It was to be filled on the basis of performance on a civil service examination, and Maunder scored high enough to get the job. He was assigned to photograph sunspots and measure their areas and positions on the solar disk. In the course of this work he became impressed with the fact that "the smallest portion of the Sun's surface visible by us as a separate entity, even as a mathematical point, is yet really a wide extended area."[11] In an article titled "The Tenuity of the Sun's Surroundings," which was published in *Knowledge* on March 1, 1894, Maunder expanded on this point:

> Now this fact has an important bearing on some of our theories. We easily fall into the mistake of supposing that the most delicate details which we can see really form the ultimate structure of the solar surface; but it is not possible that they can do so. The finest granule, the smallest pore, as we see it, is only the integration of a vast aggregation of details far too delicate for us to detect; and the minute speck of brighter or duller material may, and probably does, contain within itself a wide range of brilliancy, not to speak of varieties of temperature, of pressure, of motion, and of chemical composition.[12]

Later, Maunder applied the same argument to the surface of Mars, writing: "We have no right to assume, and yet we do habitually assume, that our telescopes reveal to us the ultimate structure of the planet."[13] When he performed experiments to determine how small a spot made with India ink on white glazed paper and viewed in dull, diffused daylight could be detected without optical assistance, he found that the limit of his vision for a circular spot proved to be 30–36″ of arc. A spot of 20″ was quite invisible; one of 40″ was distinctly seen. Much to his surprise, however, the limit for a straight line proved to be only 7″ or 8″, while a chain of dots, each of 20″ diameter and separated by an interval three times as great, was easily seen as a continuous straight line. "In each case," he noted, "the object was unmistakably *discerned,* and appeared as a line or dot; it was not, of course, *defined* so as to be seen in its true form." He concluded that

> the rough little experiments to which I have alluded may, I think, throw some light on the "canal system."... [A] narrow dark line *can* be seen when its breadth is far less than the diameter of the smallest visible dot. Further, a line of detached dots will produce the impression of a continuous line, if the dots be too small or too close together for separate vision. There are some intimations that this may be the next phase of the "canal" question, Mr. Gale, of Paddington, New South Wales, having broken up one "canal" into a chain of "lakes" on a night of superb definition, Mars being near the zenith, and Prof. W. H. Pickering, at Arequipa, having under equally favourable circumstances detected a vast number of small "lakes" in the general structure of the "canal system."[14]

At the moment, however, the doubters were still in the minority. The momentum was still on Lowell's side, fed by the endlessly fascinating idea of intelligent life on another world to which it seemed that he had given definitive form and which appeared to be fast on the way to being accorded, in William Graves Hoyt's phrase, "conditional credulity."[15] Some of Lowell's observations, it is true, were soon to be called into serious question, but this involved Venus, not Mars.

THOUGH LOWELL was later to claim that at Flagstaff, "details invisible at the average observatory were presented at times with copper-plate distinctness, and, what is as vital, the markings were seen hour by hour, day by day, month by month,"[16] after he left in November 1894 for Boston and the European tour, conditions at Flagstaff took a turn for the worse. During the winter of 1894–95, A. E. Douglass found "not a single perfect night . . . and scarcely one or two which could be called good,"[17] so Lowell decided to cast around for a more favorable climate from which to observe Mars at its next opposition, on December 1, 1896. (While he was abroad in 1895 he is said to have looked closely at several alternative sites, including Pic du Midi in southern France, which, ironically, would later become the site of an observatory famed for its seeing, and the Sahara, where conditions were mediocre at best.)

Meanwhile, he had acquired at a cost of $20,000 a permanent telescope for his observatory—a 24-inch (61-cm) refractor made by Alvan Clark's son, Alvan Graham Clark. It was installed in Flagstaff in August 1896. Since Mars that summer showed a disk only 8″ of arc across, Lowell, who by now was back at Flagstaff, tested his new telescope on the inner planets, Mercury and Venus. What he found, though not directly relevant to Mars, had important consequences for the credibility of his Mars observations, and thus deserves at least a brief digression.

Schiaparelli, as we have seen, had assigned a captured rotation to both Mercury and Venus. With his new refractor Lowell found the markings on Mercury so unmistakable that within a day or two he was confident that Schiaparelli had been correct. Lowell also tackled Venus, which despite its brilliance had generally revealed only vague, elusive markings even to the best observers, including Schiaparelli himself. Unexpectedly, Lowell found "many markings" on the nearly full disk.

Lowell rushed into print with his Venus observations, just as he had previously done with his "facts" about Mars.[18] His basic description of the planet was unlike anything reported by previous observers. The markings he found there were, he said, "in the matter of contrast as accentuated, in good seeing, as the markings on the Moon and owing to their character much easier to draw. . . . They are rather lines than spots. . . . A large number of them, but by no means all, radiate like spokes from a certain center."[19] In late 1896, Lowell moved his telescope and observatory to Tacubaya, near Mexico City, where he hoped to find more favorable conditions for the Martian opposition. Mars produced no new sensations, but the Venusian spoke system continued to hold forth, and Lowell regarded his observations as furnishing final proof of Schiaparelli's earlier tentative announcement that the planet always holds the same face toward the Sun. The observations, he wrote, "may be said to have put the rotation period beyond even reasonable doubt."[20]

Instead of acquiescence to his views, Lowell found on his return from Mexico in April 1897 that they had met with almost universal incredulity. E. M. Antoniadi, for instance, criticized those who "forgetting that Venus is decently clad in a dense atmospheric mantle, cover what they call the 'surface' of the unfortunate planet with the fashionable canal network."[21] Meanwhile, Lowell had decided to move the observatory back to Flagstaff, convinced that conditions were not so bad there after all. He had already established his routine of using a diaphragm with his large refractor in order to eliminate the blurring effects of atmospheric eddies swirling overhead. Generally, he found that with a diaphragm cutting the aperture to 12 to 16 inches, "detail which would remain hopelessly hid with the full aperture . . . starts forth to sight."[22] This concept was actually less foolish than it may sound; the effects Lowell described are quite real, and they do blur the image in a large telescope.[23] Certainly at the time, with the exception of Barnard's after-sunrise observations with the Lick refractor in 1894, large telescopes had yet to demonstrate their effectiveness in planetary work, and on the whole Lowell's ideas about seeing, partly drawn from Pickering's and especially Douglass's investigations, were well in advance of most of his contemporaries.

At this point Lowell suffered "a breakdown of the nerves." He was forced to retreat from astronomy for the next four years, and Douglass became acting director at the observatory. Thus far, he had yet to emerge fully from the shadows of Pickering and Lowell.[24] His first priority was to attempt to defend Lowell's observations of Venus and Mars, but he also discovered, in his own right, a set of linelike markings on the Galilean satellites of Jupiter. The markings were immediately criticized by Barnard, who had failed to find anything similar with the refractor at Lick. Douglass's own critical faculties were now aroused, and he began experimenting with artificial planet disks. His confidence was shaken when he found that such disks viewed from a distance of about a mile appeared to have illusory markings that bore an uncomfortable resemblance to some of those reported as occurring on Venus, such as dark faint shadings and cusp caps.[25] After Lowell returned to Flagstaff in spring 1901, Douglass rather indiscreetly wrote to Lowell's brother-in-law, William Lowell Putnam, who had served as trustee of the observatory during Lowell's illness, to complain that Lowell's method was "unscientific" and consisted of "hunt[ing] up a few facts in support of some speculation."[26] Eventually Lowell learned of the letter, and Douglass was summarily dismissed.

After Douglass left—he went on to found the Steward Observatory at the University of Arizona in Tucson and discovered a lifelong fascination for tree rings—Lowell was briefly left without assistants.[27] Before the year was out, however, he was joined first by Vesto Melvin Slipher (Lowell set him to work with a new spectrograph in order to determine the rotation of Venus), and then by Carl Otto Lampland and Vesto's younger brother, Earl Carl Slipher. Apparently Douglass's artificial planet experiments had shaken even Lowell's confidence in the reality of the spokelike markings on Venus, for briefly, in 1902, he published a retraction.[28] However, on returning to the telescope in 1903, he found the same spokelike markings staring back at him, "with a definition to convince the beholder of an objectiveness beyond the possibility of illusion."[29] He remained convinced of their reality for the rest of his life,[30] but in later years he expended relatively little effort in publicly defending them, for he faced a mounting challenge to a much more cherished belief—the reality of the Martian canals themselves.

THE YEAR that Lowell recovered the spoke system on Venus, thereby going against the grain of most observers of the planet before and since, Maunder and J. E. Evans made a significant contribution to the canal debate with their paper entitled "Experiments as to the Actuality of the 'Canals' of Mars." The paper summarizes the results of an experiment in which boys at the Royal Greenwich Hospital school were asked to reproduce a disk on which no canals had been drawn but only "minute dot-like markings." Maunder and Evans found that when the disk was viewed from a certain distance, the boys drew "canals."[31] Lowell, as might be expected, was unimpressed by the "small boy theory" and argued that the canal question ought to be decided not by experiments but by "actual observation directed to that end," expressing confidence that "if England would only send out an expedition to steady air . . . it would soon convince itself of these realities."[32]

In fact, evidence was mounting that the Martian surface did indeed consist of complex structures lying generally just below the threshold of distinct perception with ordinary telescopes. Among the most remarkable observations in this regard are those made by Percy Braybrooke Molesworth, a captain in the British army stationed at Trincomalee, Ceylon, within only a few degrees of the equator. In 1896, using a 9.25-inch (23-cm) reflector, Molesworth saw more canals than any other member of the British Astronomical Association, but he found them to be broad and curving and entirely lacking in "the hard line-like appearance with which they are drawn by Schiaparelli."[33] When he was able to obtain better observations of Mars with a 12.5-inch (32-cm) reflector equipped with a clock drive, in 1903, Molesworth reported that "the amount of detail is bewildering, and I despair of giving even an approximate idea of it in a drawing." Again: "The broad effects one draws are simply the combined results of myriads of small details, too minute to be appreciated separately. . . . I cannot help being certain that our present instruments are quite incapable of dealing with the details of Mars, and that even the best and most careful drawings give an utterly wrong idea of the configuration of his surface. The eye interprets as well as it can, but the task is beyond its power."[34] The light and dark areas on Mars appeared to differ only in the tone of the background, since "under the best conditions the maria break up completely. There is no regular shading . . . only a confused mass of streaks, splashes and stipplings of various tones."[35]

Perhaps the most vigorous advocate of the position that the canals were an illusion produced by smaller unresolved features was the Italian observer Vincenzo Cerulli,[36] a man who had studied in Berlin and Bonn, and for a short time had served as assistant at the observatory of the Collegio Romano. Like Lowell, he was a man of independent means, and in 1890 he constructed his own observatory near Teramo, Italy, on a hilltop which he named Collurania (Urania Hill). He equipped his observatory with a fine 15.5-inch (39-cm) Cooke refractor, the largest instrument in Italy after the 19-inch (49-cm) refractor of the Brera Observatory in Milan. Like Flammarion's observatory at Juvisy and Lowell's at Flagstaff, the Collurania Observatory was a temple to the sky which owed its existence primarily to its founder's fascination with the planet Mars.

The turning point in Cerulli's study of Mars came on January 4, 1897, when there were "some moments of perfect definition [in which] Mars appeared perfectly free from undulation." Under these conditions, Cerulli watched with astonishment as the canal Lethes "lost its form of a line and altered itself into a complex and indecipherable system of minute patches."[37] Though he had previously recorded, and indeed would continue to record, numerous canals, not only in the light areas but in the dark areas as well, Cerulli henceforth regarded the whole network with suspicion and published a book on the subject, *Marte nel 1896–97*. These observations were especially significant because Schiaparelli looked to Cerulli as his most promising successor, confiding to Otto Struve: "I had once hoped that such work might be done by Mr. Percival Lowell, but he is gravely ill; moreover he is more a literary man than an astronomer, much attracted to theatrical matters and sensational news. Signor Cerulli at Teramo is a man to be taken more seriously, and may yet do something solid, if only he resists the tendency he has shown to judge the observations by the theories in his head."[38]

Needless to say, Schiaparelli did not share Cerulli's interpretation of the canals.

LOWELL, seized with a new intensity following his return to his observatory after his illness, spent the opposition years of 1901 and 1903 making extensive visual observations, despite the fact that Mars's distance from the Earth made these unfavorable opportunities. Among other things, he plotted the visibility of the canals as a function of Martian dates and latitudes to produce diagrams that he called "cartouches." The cartouches showed that both the canals and the dark areas participated in what he referred to as a "wave of darkening." Every spring and summer, this wave swept across the dark areas toward the equator. "Quickened by the water let loose on the melting of the polar cap," he wrote, the dark areas

> rise rapidly to prominence, to stay so for some months, and then slowly proceed to die out again. Each in turn is thus affected. . . . One after another each zone in order is reached and traversed, till even the equator is crossed, and the advance invades the territory of the other side. Following in its steps, afar, comes the slower wane. But already, from the other cap, has started an impulse of like character that sweeps reversely back again, travelling northward as the first went south. Twice each Martian year is the main body of the planet traversed by these waves of vegetal awakening, grandly oblivious to everything but their own advance.[39]

Lowell, of course, explained these widespread changes as resulting from seasonal cycles of vegetative growth and decline, but he pointed out that the vegetative revival on Mars each spring was different from revegetation on Earth. On Earth, plant growth begins in the low latitudes, where it is warmest, and then progresses to the higher latitudes. On Mars, he said, the revival starts nearest the poles because moisture rather than temperature is the critical factor, and this becomes available first in the circumpolar regions where the main supply is concentrated.

Apart from the wave of darkening, Lowell noted changes in color that also seemed to be seasonal in nature and lent still further support to the vegetation theory. In 1903 he found that the Mare Erythraeum had turned from blue-green to chocolate brown as Martian autumn gave way to winter, then gradually began to become green again as Martian spring approached. On observing the same changes again two years later, he wrote: "Unlike the ochre of the light regions generally, which suggest desert pure and simple, the chocolate-brown precisely mimicked the complexion of fallow ground. When we consider the vegetal-like blue-green that it replaced, and remember further the time of year at which it occurred in both these Martian years, we can hardly resist the conclusion that it was something very like fallow field that was there uncovered to our view." [40]

Lowell was joined in his campaign to prove that there was life on Mars by his assistants V. M. Slipher and Lampland. In 1903, Slipher began spectrographic observations of Mars in an attempt to refute Campbell's negative observations and demonstrate the presence of water vapor there. Unable to obtain plates sensitive to the red part of the spectrum where the water vapor lines are found, he was forced, at least temporarily, to admit defeat. Lampland's project was photography of the planet, and in 1905 Lowell announced that his assistant had succeeded in photographing some of the canals, an achievement for which he was awarded the medal of the Royal Photographic Society. Schiaparelli wrote to Lowell, "I should never have believed it possible." [41]

Based mainly on his observations of 1903, though supplemented with results from the 1905 opposition, which he also observed extensively, Lowell wrote *Mars and Its Canals,* his magnum opus on the red planet. The book appeared in December 1906, at a time when public interest was intense. In addition to providing a massive résumé of Lowell's visual studies of the planet and his controversial interpretations, the book makes brief mention of Lampland's photographs, noting that "thus did the canals at last speak for their own reality themselves." [42] Unfortunately, Lowell was unable to find a way to adequately reproduce Lampland's tiny, delicate images, each only a quarter inch (6 mm) across, and none of the photographs appear in the book. Among the various experts who examined the original images, some agreed that they showed the canals, but others were not so sure. Inevitably someone pointed out that even though there might *appear* to be linear details in the photographs, it did not follow that this was the actual form of the features on the planet's surface.[43]

The most notable response to the publication of *Mars and Its Canals* came from Alfred Russel Wallace, the octogenarian naturalist who as a young man had been co-discoverer (with Charles Darwin) of the theory of evolution by natural selection. Asked to review Lowell's book, Wallace, who regarded Lowell's book as "a challenge, not so much to astronomers as to the educated world at large," responded with a remarkable book of his own, *Is Mars Habitable?* In it he launched a devastating critique, not of Lowell's observations, which he accepted, but of Lowell's conclusions from them.[44] Lowell had argued in a paper published in 1907 that the temperature on Mars was "comfortable as the south of England."[45] Wallace effectively impeached Lowell's estimate of the albedo (reflectivity) of the Earth, which had been crucial to his reasoning.[46] Instead of comfortable warmth, Wallace surmised, the temperature almost everywhere on Mars was probably very far below the freezing point of water. If so, the polar caps might well be frozen carbon dioxide, as had been proposed some years earlier by A. Cowper Ranyard and G. Johnstone Stoney, rather than water ice. As for Lowell's idea of water-filled canals, Wallace wrote scathingly: "Any attempt to make that scanty surplus, by means of overflowing canals, travel across the equator into the opposite hemisphere, through such terrible desert regions and exposed to such a cloudless sky as Mr. Lowell describes, would be the work of a body of madmen rather than of intelligent beings. It may be safely asserted that not one drop of water would escape evaporation or insoak at even a hundred miles from its source."[47]

Despite Wallace's book, however, the public continued to support Lowell and his theory of life on Mars. His secretary, Wrexie Leonard, described the frenzied response to his 1906 lecture series for the Lowell Institute at Boston's thousand-seat Huntington Hall: "Standing room was nil, and demands for admission were so numerous and insistent that repetitions were arranged for the evenings. At these repeated lectures the streets near by were filled with motors and carriages as if it were grand opera night!"[48]

Lowell remained very much in the public eye at the opposition of 1907, when Mars was near the Earth but too far south to be studied satisfactorily from northern observatories. While he remained in Flagstaff to observe the planet visually with the 24-inch refractor, his assistant E. C. Slipher went with Amherst College professor David Peck Todd to Alianza, Chile, in order to photograph the planet with Amherst's 18-inch (46-cm) refractor, which had been shipped to South America specifically for that purpose. The telescope was set up in the open desert with only the sky for a dome, and some of Slipher's 13,000 images were alleged to have captured canals—including, Slipher announced, some of the more prominent doubles. Naturally, Lowell was excited by this apparent success, and he wrote to Todd: "Bravo! . . . The world, to judge from the English and American papers, is on the *qui vive* about the expedition, as well as about Mars. They send me cables at their own extravagant expense and mention vague but huge (or they won't get 'em) sums for exclusive magazine publication of the photographs."[49] Eventually the bidding war for publication rights was won by *Century* magazine, but as before, the delicate detail of the photographs proved difficult to reproduce satisfactorily, and the photographic evidence of the canals remained inconclusive.

ALL OF this controversy only served to pique interest in the next opposition, in 1909. Mars would be only slightly closer to the Earth than it had been in 1907, but it would be far better placed for northern observatories. More large telescopes would be trained on it than at any other time in history, thus setting the stage for what would prove to be the dramatic climax to the long and heated debate over the Martian canals.

Opposition 1909

In July 1907, G. V. Schiaparelli wrote a remarkable letter to Vincenzo Cerulli in which he considered the view of a printed page from various distances. He noted that in a first stage, A, the vision is confused and the page appears as a gray square; at a next stage, B, this view is replaced with a vision of geometrical lines; at a third stage, C, one suspects breaks and irregularities; finally, at stage D, one is able to read the individual letters. The relevance of this to the observation of Mars, Schiaparelli wrote, is that

> the first observers of Mars, to 1860, lived in stage A. Since this epoch, Secchi, Kaiser, and Dawes came near to stage B, finding some lines. . . . In the years after 1877 the view produced in me and others was stage B—a vision apparently complete and accurate of single and double lines on the planet. Now, thanks to you [Cerulli], we are entering stage C; the naive faith in the regularity of the lines is shaken, and we have the prospect of yet another stage, D, in which the appearance of lines will resolve into forms of a different order—closer to the true structure of the Martian surface. But will this, then, be the final truth? No; for of course as optics continue to improve, the process will proceed to other stages of vision, or illusion. My thanks to you for the progress you have realized along this stairway.[1]

It is hard to believe that anyone capable of such penetrating analysis would retreat again into the realm of error and illusion, and yet Schiaparelli did just that after studying Lowell's 1907 photographs of Mars. So, apparently, did Cerulli himself. Schiaparelli wrote to E. M. Antoniadi in 1909: "The polygonations and geminations for which you show so much horror (and, with you, so many others) are a proved fact, against which it is needless to dissent. Dr. Cerulli was convinced some weeks ago. I have shown him a series of fine photographs obtained by Mr. Lowell in July, 1907; he was able to see the doubling of the Gehon, the Ganges and several others."[2]

Schiaparelli was now an old man, but he was still active; he observed the planet at the next opposition of 1909. In June 1910, however, he suffered a stroke, and he died in Milan on July 4. His last recorded utterances about Mars, in May 1910, show that he had returned to his long-held views. "I am of the opinion," he summed up, "that the geometrical and regular lines (the existence of which is still disputed by many) teach us nothing at present in regard to the probability or improbability of intelligent beings on the planet. However it would be worthwhile were someone to collect everything . . . that can reasonably be said on the subject. And from this viewpoint, I hold in high esteem the noble-minded endeavours of Mr. Lowell . . . as well as his very perceptive arguments on the matter."[3] He thus died, as he had for so long lived, an agnostic concerning the meaning of the canals and the question of Martian life, though toward the end he was leaning heavily toward Lowell's views.

THOUGH THEY had nipped Schiaparelli's skepticism in the bud, Lowell's photographs failed to overcome the hardened opposition of W. W. Campbell, Lowell's longtime nemesis and now the director of the Lick Observatory. Campbell wrote to George Ellery Hale at Mount Wilson:

You have of course noticed that Lowell, the past year or two, has been making much ado in public, and in many matters quite unprofessionally. I have occasionally thought of putting my finger publicly on the weak points, but have serious doubts as to the usefulness of such an unpleasant undertaking. I do not believe that either his photographs or Todd's record any markings on Mars that have not been conceded to exist by experienced observers of the planet for twenty-five years past; and which can be seen with a six-inch telescope better than they have been photographed. . . . I think Lowell and Todd are going to be a trial to sane astronomers. . . . My question is just how far they should be allowed to go before somebody steps on their rope.[4]

In addition to his photographic evidence of canals, Lowell was also touting the spectrographic discovery of water vapor on the planet. During January 1908 the air at Flagstaff had been exceptionally dry, and V. M. Slipher had obtained photographs of the spectrum of Mars which seemed to indicate that the "a" band indicating water vapor (near wavelength 7,150) was strengthened relative to that in the Moon's spectrum. Lowell sent copies of Slipher's plates to the *Encyclopaedia Britannica,* telling Slipher that there "they are sure to go 'thundering down the ages'—as Prof. [E. S.] Morse would say—bumping at every obstruction."[5] But Campbell, who had found no definite evidence of water vapor from Mount Hamilton in 1894, was skeptical of this pronouncement as well, telling Hale: "From the first, I have had no confidence in their reported evidence of water vapor, and these spectrograms do not change my opinion. The critical band lies just in the beginning of the region where Slipher's plates fall off exceedingly rapidly in sensitiveness, and we all know very well that in such regions the apparent contrasts may vary widely from the truth, both on the original negatives and the photographic copies."[6] Campbell himself carried out an arduous expedition to the summit of Mount Whitney (elevation 14,495 ft, or 4,418 m) in late August and early September 1909 in order to obtain spectrograms with Mars close to the Earth and next to the Moon in the sky. The result was negative, thus confirming his 1894 observations. Campbell did not claim that there was no water vapor in the Martian atmosphere at all, only that "any water vapor . . . must have been much less extensive than was contained in the rarified and dry air strata above Mount Whitney."[7] His result would stand the test of time; not until 1963 were very minute traces of water definitively detected in the Martian atmosphere.[8]

As in 1894, Campbell's negative spectroscopic observations in 1909 created relatively few headlines, at least compared with the visual observations of the planet, which again grabbed most of the attention. As Mars again approached the Earth, Lowell had hoped to have a larger telescope to turn toward it—at least in part to combat the allegation that no large telescope had ever shown the canals. At first he had considered buying another refractor, possibly as large as 50 inches (1.27 m), but the cost estimates proved sobering even to Lowell, and he began "reflecting on the reflector"—this despite having only a few years before declared that "reflectors should be shunned; alluring though they be. . . . [F]or planetary detail they are well-nigh worthless."[9] He discussed plans for an 84-inch (2.14-m) reflector with George W. Ritchey, the brilliant telescope maker then at Mount Wilson, but finally contracted for a more modest 40-inch (1.02-m) reflector with the Clark firm, at that time headed by Carl A. R. Lundin.[10] But Lowell had reflected too long, and the telescope was not ready in time for the September 1909 opposition. In any case, it failed to perform up to expectations, largely because of disastrous seeing brought about by Lowell's novel but ill-advised decision to mount it below ground level.

Lowell was not the only astronomer who hoped to turn a large telescope on Mars in 1909. Eugène Michael Antoniadi (fig. 15), who used the "Grand Lunette"—the 33-inch (83-cm) Henry refractor at Meudon Observatory—to obtain some of the most remarkable views of the planet ever obtained in the prespacecraft era, was another.

Of Greek descent, Antoniadi was born in Constantinople (now Istanbul) in 1870. His interest in astronomy was awakened early. By his late teens he was already observing with a 3-inch (76-mm) refractor at Constantinople and on the island of Prinkipio in the Marmara Sea, and had begun submitting his drawings to Camille Flammarion's Société Astronomique de France and the British Astronomical Association. Antoniadi was a remarkably talented draftsman, and his drawings attracted immediate attention; Flammarion was so impressed that he invited Antoniadi to visit him at Juvisy in 1893, and promptly hired him as an assistant.[11]

For a modest salary of 300 francs a month, Antoniadi was expected to observe six nights a week, usually with Flammarion's fine 9-inch (23-cm) Bardou refractor, and to copy all his observations into the official records of the observatory. Antoniadi would leave his residence in Paris, travel to Meudon, and ascend to the observatory by means of a tower so that he could reach the dome without having to pass through Flammarion's private quarters. Mars was

FIGURE 15. E. M. Antoniadi, on board the ship *Norse King* during an 1896 eclipse expedition. (Courtesy Dr. Richard McKim and Council of the British Astronomical Association)

the chief interest of both men. Flammarion, who had just published the first volume of his classic *La Planète Mars,* regarded himself as the greatest living authority on the planet, and in 1896 Antoniadi had become the director of the Mars Section of the British Astronomical Association (BAA).

In the 1890s, Antoniadi collaborated with Flammarion on several canal-filled maps of the planet (fig. 16). At this time he was convinced of the canals' objective reality. However, his successive Mars Section reports to the BAA show a gradual change in viewpoint. In his report for 1896–97, he noted that "the canals were seen by all the working members of the section invariably," [12] with Antoniadi himself, who sighted 46 canals, ranking as the third most prolific canalist behind Percy Molesworth and Rev. T. E. R. Phillips. Phillips, observing for the first time with a 9.5-inch (24-cm) reflector, went so far as to say: "My experience of Martian observation this winter has led me to believe that Mars is not nearly so difficult an object as is commonly supposed, and that many of the canals are easy." [13] Antoniadi's experience was somewhat different; at Juvisy he found that "the canals are very difficult objects, visible only in rare glimpses." He added that "but for Prof. Schiaparelli's wonderful discoveries, and the foreknowledge that 'the canals are there'," he would have missed three quarters at least of those seen. [14] Nevertheless, though convinced in general of the canals' objective existence, he already found it prudent to throw out the work of at least one embarrassingly prolific enthusiast, C. Roberts, who despite using only a 6.5-inch (16-cm) reflector recorded no less than 134 canals on his highly stylized drawings. Antoniadi explained: "It was thought safer to avoid introducing uncertain data in the general excellence of the section's work, and not overcrowd our already crowded chart with the most daedelian network ever devised." [15] Incidentally, even Roberts's efforts were surpassed by those of the quixotic Serbian astronomer Spiridion Gopčević, or Leo Brenner as he called himself, whose 1896–97 map based on observations made with a 7-inch (18-cm) refractor shows 164 canals, including no less than 18 radiating from the small dark spot called Trivium Charontis. Moreover, Brenner claimed to have detected 34 canals using a tiny 3-inch (76-mm) refractor! [16]

Antoniadi's next report (1901), on the results of the 1898–99 opposition, includes the following comment: "Notwithstanding

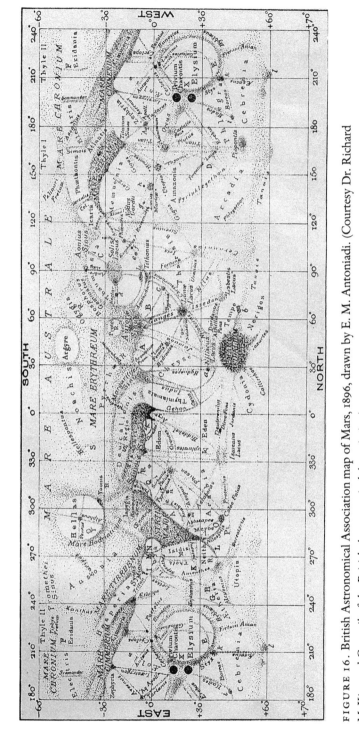

FIGURE 16. British Astronomical Association map of Mars, 1896, drawn by E. M. Antoniadi. (Courtesy Dr. Richard McKim and Council of the British Astronomical Association)

the natural skepticism of many scientific men, every opposition brings with it its own contingent of confirmations of Schiaparelli's discovery of linear markings, apparently furrowing the surface of the planet Mars. The difference between objective and subjective in the daedelian phenomena presented by these appearances will be the work of future generations. But the value of the great Italian results will be everlasting." [17] His qualifications were justified, for by now he had become convinced that the geminations—certainly the most sensational of the "great Italian results"—were illusions, the result of focusing errors or mere eye fatigue. (He later retracted this particular theory, but there has never been much doubt that they were an optical effect, apparently involving some kind of astigmatism in the observer's eye—I must admit that I have never seen a fully convincing explanation.) Moreover, his confidence in the whole network had been badly shaken by the "discovery" by Lowell and his assistants of what Antoniadi referred to as "subjective" linear markings on Mercury, Venus, and the Jovian satellites. Whereas in 1898 Antoniadi had stated that "despite the skepticism of several eminent authorities, I do not hesitate to say that the famous canals of Mars have a true objective existence," by 1902 he characterized his position as "agnostic." [18]

The decisive event had come with his break with Flammarion in that same year, 1902. Antoniadi had just married Katharine Sevastupulo, who belonged to one of the leading families in Paris's Greek community. She may have had independent means. In any case, the marriage was good for Antoniadi; he had frequently suffered from poor health toward the end of his association with Flammarion, but after his marriage his health began to improve. (One suspects that some of his complaints, at least, were psychosomatic; Maunder's wife, Annie, complained that Antoniadi worried too much.) [19] For a while Antoniadi considered becoming an Englishman, but at last he decided to remain in France; he and Katharine found an apartment on the Rue Jouffroy, in one of the most expensive districts in Paris, and there they remained for many years.

One detects a sea-change in Antoniadi's view of the Martian canals after his departure from Juvisy. He still hesitated to place them into the same category as the disputed Lowellian markings on Venus and Mercury, pointing out that "the hard line-likeness of the 'canals' is almost sure to be experienced by painstaking observers of the planet; and this circumstance cannot be treated lightly as illusive."[20] Nevertheless, in order to avoid "the possibility of our representation of Mars [being] profaned by doubt," Antoniadi took a decisive step in his Mars Section report on the 1903 opposition.[21] In addition to the customary canal-filled chart that provided a summary of all submitted observations, he prepared another version from which all the canals had been carefully expunged. Maunder wrote of this radical departure: "We seem to have returned to the pre-Schiaparellian age. . . . Is it a retrogression or an advance? . . . Either way we may take it as marking an epoch; for it is practically the first time for five-and-twenty years that a chart of Mars has appeared in which the canal-system was not predominant. Even should it be condemned as unscientific, it would still have an historic importance as marking the growing strength of a reaction."[22]

Over the next several years Antoniadi gave up much of his astronomical work, instead concentrating on an intensive study of the architecture of the Mosque of Saint Sophia in Constantinople that eventually led to the publication of a three-volume work on the subject. Not until 1909 were his abilities as an observer and astronomical artist fully reawakened, when he appeared on the stage to play the greatest role of his life. Henri Deslandres, the director of the Meudon Observatory, placed the Grand Lunette—then, as now, the largest refractor in Europe, and the third largest in the world—at his disposal for the favorable opposition.

Even before Antoniadi was given this great opportunity, he was already studying Mars with his own 8.5-inch (22-cm) reflector. In August, he found the markings on the planet unusually pale, an aspect he attributed to a veil of "pale lemon" haze. "As far as the outlines of the markings are concerned," he added, "Mr. Lowell's maps are the most accurate ever published."[23] Antoniadi wrote to Lowell shortly before he started work with the Grand Lunette, and by return post received some advice. "I am glad that you are to use the Meudon refractor," Lowell said, but then enjoined him to "remember that you will have to diaphragm it down to get the finest details. Even here we find 12 to 18 inches [30–46 cm] the best sizes."[24]

At Flagstaff, Lowell had found it generally necessary to stop down his 24-inch refractor by means of a diaphragm, usually to 12 to 16 inches. This, he maintained, was because of the presence of eddies in the air, which caused blurring of the images when the full aperture was used. Indeed, he had a valid point; these eddies do exist, and they vary in size from millimeters to tens and even hundreds of meters—the largest of them naturally produce the greatest variations in refractive index, and since a larger telescope samples larger and stronger areas of turbulence, the blurring that Lowell wrote about is quite real.[25] Moreover, there was another advantage to using a diaphragm. All large lenses, including Lowell's own, which has perhaps the finest optical figure of any lens ever made, suffer seriously from chromatic aberration. This causes a planet as seen against a dark sky background to be surrounded with an obnoxious magenta or purplish haze, which is highly deleterious to the perception of fine details. Diaphragming reduces this bothersome effect. I have used the Lowell refractor myself, and based on my experience suspect that the reduction of chromatic aberration, rather than the fact that Lowell was viewing through a narrower, less turbulent light path, accounts for most of the advantage he found in stopping down to 12 to 16 inches.

Before this letter arrived, Antoniadi had already had his first chance with the Grand Lunette, which does not seem to have been equipped with a diaphragm at the time. Though Antoniadi had earlier conceded the superiority of the "ideal definition" at Flagstaff, in practice he found conditions at Meudon more favorable than he had expected. The building housing the Grand Lunette (fig. 17) stands on the edge of a high terrace; to the east there is a sheer drop to the Meudon Park below, so that the seeing is generally very good for objects to the east of the meridian—even objects that are low and rising. To the west there is no such drop, and air currents from the ground exert more deleterious effects, with the result that the seeing often deteriorates rapidly as objects approach the meridian. Under the conditions most apt to bring favorable seeing—with Mars to the east and rising—Antoniadi, on September 20, 1909, made his first observations with the Grand Lunette. There was a temperature inversion that night over Paris; the air had arranged itself into stratified layers that remained stable for seven hours, allowing images of Mars that were simply glorious. It is supremely ironic that Antoniadi's first views of the planet with the large instrument were to prove the best of his career. "The first glance cast at the planet on September 20 was a revelation," he wrote. "The planet appeared covered with a vast and incredible amount of detail held steadily,

FIGURE 17. The Meudon Observatory near Paris.
(Photograph by William Sheehan, 1975)

all natural and logical, irregular and chequered, from which geome-
try was conspicuous by its complete absence." A "maze of complex
markings" covered the south part of Syrtis Major, which was then
approaching the central meridian; the deserts of Libya and Hesperia appeared shaded, and Mare Tyrrhenum looked "like a leopard
skin."[26] He described the land between Syrtis Major and Hellas as
being "like a green meadow, sprinkled with tiny white spots of various sizes, and diversified with darker or lighter shades of green."[27]
He later told Lowell that after he got over his initial excitement at
the "bewildering" amount of detail visible, "I sat down and drew
correctly both with regard to form and intensity all the markings
visible. . . . However, one third of the minute features I could not
draw; the task being beyond my means"(fig. 18).[28]

Five days after these breathtaking views of the planet, Antoniadi

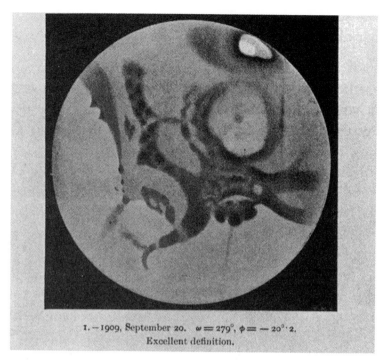

I. – 1909, September 20. $\omega = 279°$, $\phi = -20°\cdot2$.
Excellent definition.

FIGURE 18. A drawing of Mars by E. M. Antoniadi, September 20, 1909, showing the region centered on Syrtis Major; it was made on Antoniadi's first night of observing with the great Meudon refractor. (Courtesy Dr. Richard McKim and Council of the British Astronomical Association)

wrote excitedly to W. H. Wesley, secretary of the Royal Astronomical Society, "I have seen Mars more detailed than ever, and I pronounce the general configuration of the planet to be very irregular, and shaded with markings of every degree of darkness. Mars appeared in the giant telescope very much like the Moon, or even like the aspect of the Earth's surface such as I saw it in 1900 from a balloon at a great height (12,000 feet)."[29] He continued to observe and draw the planet. He pointed out to Lowell that "the tremendous difficulty was not to *see* the detail, but accurately to *represent* it. There, my experience in drawing proved of immense assistance."[30] Again he emphasized that the detail he was seeing was not in the least geometric; it was highly irregular and completely natural in appearance. "Bewildering" was the word he would use again and again in describing it.

Antoniadi made all of these observations using the full aperture of the 33-inch refractor. In Lowell's view, this was a problem, and he somewhat perversely dubbed "best" the one drawing Antoniadi had marked "tremulous definition." This drawing showed canals, which Antoniadi insisted appeared only by glimpses when the seeing was poor. The other drawings showed the planet as it had looked in seeing that Antoniadi ranked as "moderate," "splendid," and "glorious," but Lowell dismissed these as "not so well defined," and he reminded Antoniadi once again of the blurring problem with large telescopes: "This is the great danger with a large aperture—a seeming superbness of image when in fact there is a fine imperceptible blurring which transforms the detail really continuous into apparent patches. On the other hand, a bodily movement often coincides with the revelation of fine detail. The subject we have carefully investigated here and all of our observers recognize it."[31] Antoniadi politely but firmly disagreed: "I understand from your letter that you consider my knotted Mare Tyrrhenum as due to blurring; but I beg to call your attention to the fact that I was holding steadily this knotted structure; and that two days ago, I found this particular knotted appearance confirmed by photography."[32] Henceforth he became a vigorous advocate of the crushing superiority of large telescopes for visual planetary work.[33]

Antoniadi was a skilled observer. On the other hand, the Meudon telescope suffers from chromatic aberration no less than the Lowell instrument—indeed more so, on account of its greater size. In such an instrument—and this I can say from personal experience—the image of a bright planet like Mars is awash in an obnoxious haze of magenta or violet light. Lowell's suggestion of a diaphragm was not as unreasonable as Antoniadi obviously thought it, though even better would have been the use of a yellow filter to absorb some of the extraneous unfocused light. And yet there is no arguing with Antoniadi's results, and here, at least, there can be no doubt: with the Meudon instrument, Antoniadi was able to make out faint tones and subtle colors on the planet that were inaccessible in smaller instruments.

In general, Antoniadi found the best seeing at Meudon on nights when it was foggy and quiet. But he enjoyed few nights with perfect seeing—he later estimated only one night in fifty. On most evenings the image was "more or less boiling," but during the less agitated moments of the boiling image the large aperture revealed more than a smaller one would have done. Indeed, good images were "preceded by a period of slight rippling of the disk, very detrimental to the detection of fine detail. The undulations would then cease suddenly, when the perfectly calm image of Mars revealed a host of bewildering details."[34]

Such was his experience on the nights of October 6 and November 9, when he enjoyed brief periods of excellent definition in which he grasped the true character of the Martian deserts. "The soil of the planet then appeared covered with a vast number of dark knots and chequered fields," he wrote, "diversified with the faintest imaginable dusky areas, and marbled with irregular, undulating filaments, the representation of which was evidently beyond the powers of any artist. There was nothing geometrical in all this, nothing artificial, the whole appearance having something overwhelmingly natural about it."[35] In the desert known as Amazonis he had held some of this detail for ten to twelve seconds at a time. Instead of the "hideous lines" that were wont to appear in brief glimpses under conditions of indifferent seeing, he made out "a maze of knotted, irregular, chequered streaks and spots." Even held thus steadily, the features were so complex that Antoniadi despaired of drawing them, though he provided Lowell an impressionistic sketch of what he had seen. There is a striking similarity between this sketch and the *Mariner 9* and Viking imagery, the resemblance of the pattern of windblown streaks around cratered terrain to Antoniadi's sketch being highly suggestive of the real structure of the surface on this part of the planet.

With regard to the canals, Antoniadi believed that, in some cases, there was at least some objective basis to their fleeting apparitions. The Jamuna, which to Schiaparelli had appeared on June 9, 1890, as a narrow line with a breadth of 0.04" of arc, appeared in the Grand Lunette as "nothing like a regular band . . . but rather the mere irregular border of a weak halftone."[36] Other canals were similarly found to consist of "winding, irregular knotted streaks, or broad irregular bands, or groups of complex shadings, or isolated dusky spots, or jagged edges of half-tones."[37] Antoniadi thus felt that his work had provided a partial vindication of Schiaparelli's observations, but he thought otherwise of the Lowellian spiderwebs—those he regarded as absolutely illusory.

AFTER THE 1909 opposition, Mars retreated through a cycle of less favorable oppositions, which finally bottomed out with that of 1916. Lowell observed the planet extensively in 1911, 1913–14, and 1916, with both his 24-inch (61-cm) refractor and his newly unveiled 40-inch reflector, but found nothing to cause him to change his views. Meanwhile, he had begun to make plans for the next series of favorable oppositions, which would come in the early 1920s. Among other projects he hoped to mount a joint expedition with the French astronomer René Jarry-Desloges to the Atacama Desert in Chile. But it was not to be. Lowell's colorful career was cut short on November 12, 1916, by an intracerebral hemorrhage. Shortly before his death, he summed up his final views about Mars:

> Since the theory of intelligent life on the planet was first enunciated 21 years ago, every new fact discovered has been found to be accordant with it. Not a single thing has been detected which it does not explain. This is really a remarkable record for a theory. It has, of course, met the fate of any new idea, which has both the fortune and the misfortune to be ahead of the times and has risen above it. New facts have but buttressed the old, while every year adds to the number of those who have seen the evidence for themselves.[38]

We now know from Earth-based charge-coupled device (CCD) images (see chapter 15) and spacecraft photographs that Antoniadi was more or less right and Lowell was wrong. There are no genuine canals on Mars. Schiaparelli, Lowell, and countless others were victims of the "Grand Illusion": under certain conditions of observation, the really complex Martian details appear as a réseau of fine lines.

Ultimately, the solution to the canal mystery belonged more to the realm of perceptual psychology than to astronomy. The canals were seen by glimpses; they were fragmentary perceptions. To quote Antoniadi, who had seen many of them, as both single and double lines, at Juvisy: "A glimpsed object is not as certain as an object held steadily, and, however self-evident or trite such a remark may be, yet it is a very important one to make."[39] The periods when a large wavefront of uniform refractive index passes across the telescope aperture typically last on the order of a fraction of a second. Thus, the best seeing often occurs by what Percival Lowell once called "revelation peeps." The effect is exactly like that produced by a tachistoscope, a device used by perceptual psychologists since the turn of the century to study what takes place during interrupted or brief perceptions. The planetary observer waits for the tachistoscope flash, and waits with an expectant mind—just as Sir Ernst Gombrich listened to weak and static-filled radio transmissions during World War II; the interpretation is of "whiffs" of information coming over the airwaves (or in this case light waves). What appeared to many observers as canals and oases was actually the brain's shorthand rendering of what the eye vouchsafed to it in glimpses obtained through telescopes with modest apertures. But there was a further stage along what Schiaparelli had described as the ascending stairway of perception, the stage Antoniadi rendered in the sketches and maps he made with the Grand Lunette. Instead of lines and dots, the characteristic markings appeared as winding, knotted, irregular bands, jagged edges of halftones, and sooty patches. In turn, Antoniadi's view of the planet was itself a tentative—and blurred—vision of what would later be revealed when spacecraft visited the planet and sent back images of the surface.

THE CANALS have been disproved, but they will never be forgotten. They will always remain an important chapter in the history of Martian exploration, not least because of the literature they inspired. Beginning with H. G. Wells's interplanetary invaders and Edgar Rice Burroughs's eerie invocations of Barsoom, the canals and the dying world they were supposedly meant to save inspired much of our early science fiction—and there, at least, they will live on.[40] Moreover, observers, especially those using small instruments, will continue to have fleeting glimpses of the canals from time to time—tantalizing Lowellian moments.

The Lingering Romance

Partly because of the heated, seemingly endless, and finally inconclusive controversies about the Martian canals, and partly because of the rise of astrophysics in the years after the First World War, planetary astronomy was long relatively neglected by professional astronomers, at least in the United States. There were a few exceptions. William H. Pickering published his "Mars reports" in *Popular Astronomy* from 1914 to 1930, based largely on his observations at Mandeville, Jamaica, with an 11-inch (28-cm) refractor and, after this instrument was returned to Harvard in 1925, a 12.5-inch (32-cm) reflector. Pickering was never short of theories to explain the canals. In Arequipa in 1892, he had proposed that they were strips of vegetation lining the banks of actual canals too small to be seen, an idea Lowell enthusiastically adopted, though Pickering himself later changed his mind in favor of the theory that the lines were vegetation growing along the border of cracks similar to the steam cracks he had observed in Hawaii. (He also, incidentally, strongly advocated the existence of clouds, vegetation, and even insects on the Moon!) Most professional astronomers paid little attention to Pickering, and he died, a bitter man, in 1938.[1]

Lowell's observatory at Flagstaff continued to be an active center for Martian research after its founder's death. (Antoniadi wrote to Gabrielle Flammarion, whom Camille married after his first wife, Sylvie, died in 1919, of the fortune Lowell had left: "We can therefore talk for a long time about canals on Mars"!)[2] Earl C. Slipher was Flagstaff's leading Mars observer; a photographic specialist, he eventually obtained more than 100,000 images covering twenty-seven oppositions between 1905 and his death in 1964. He was a great admirer of Percival Lowell and remained a firm believer in the Martian canals, whose existence, he was convinced, was proved by his photographs.[3] His map from the late 1950s was adopted by the U.S. Air Force as its official chart for use with its Mars spacecraft missions—it is covered with a Lowellian canal network!

In Europe, there were a number of active Mars observers. Undoubtedly the greatest observer during the interwar years was E. M. Antoniadi, who used the Grand Lunette at Meudon to study the planet during its oppositions between 1924 and 1941. Antoniadi never belonged to the observatory staff, and referred to himself simply as the "astronome volontaire à l'Observatoire de Meudon." His classic book *La Planète Mars,* based largely on his studies at Meudon, was published in 1930; it contains very complete descriptions of the changing surface features and will always be of great value to Mars observers (fig. 19).[4] Antoniadi died in Occupied France in 1944.[5]

Other leading European observers included the Italian Mentore Maggini, who in 1926 became director of Vincenzo Cerulli's Collurania Observatory; Georges Fournier, like Antoniadi a one-time assistant of Flammarion at Juvisy, who from 1907 worked for René Jarry-Desloges at his various observing stations in France and at Sétif, Algeria; Gerard de Vaucouleurs, who carried out detailed studies in 1939 and 1941 with the 8-inch (20-cm) refractor at the observatory founded by Julien Péridier at Le Houga; and Bernard Lyot, Henri Camichel, Audouin Dollfus, and Jean-Henri Focas, who made extensive observations with the 24-inch (61-cm) refractor at Pic du Midi Observatory in the French Pyrenees, a site renowned for the quality of its seeing. Focas, like Antoniadi of Greek descent, produced the most highly detailed map of Mars ever made by a visual observer of the planet. His map expands on Antoniadi's description of the "leopard-skin" structure of the dark areas by showing them broken up into a fine mottling of small, granular patches. Apart from the visual and photographic studies of Mars, great work was done by observers armed with spectroscopes, polarimeters, and thermocouples.

It is their results that I summarize in the present chapter. I shall have very little more to say about the canals; instead I will describe

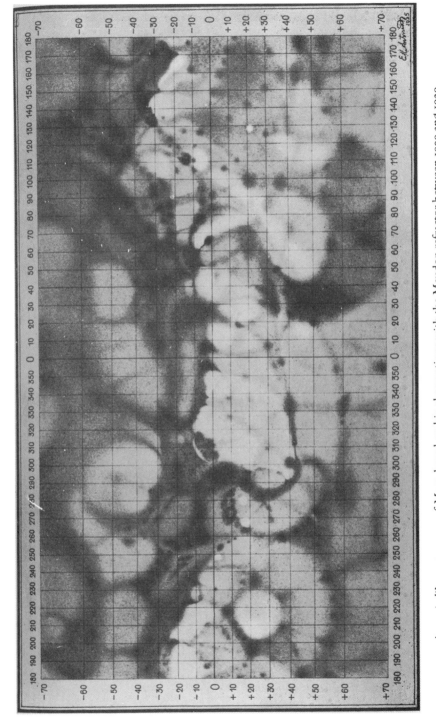

FIGURE 19. Antoniadi's master map of Mars, based on his observations with the Meudon refractor between 1909 and 1929.

the views of the planet's atmosphere, clouds, polar caps, and dark and light areas that prevailed in the years leading up to the beginning of the spacecraft era. Suffice it to say that although the canal theory had largely fallen into disrepute after Antoniadi's observations in 1909, in other respects the Lowellian synthesis remained influential. Indeed, Lowell's ideas about the water, atmosphere, and vegetation "survived and prospered. . . . [T]hese parts of Lowell's dream were repeatedly confirmed, and even embellished, by observers of Mars after 1916."[6]

ACCORDING TO the kinetic theory of gases, Mars should have a thin atmosphere. The planet, after all, is only a tenth as massive as the Earth; this means that it has a weak gravitational pull, and any gas molecules that reach a velocity of 5 km/sec (3.1 miles/sec) will escape into space. Mars thus was expected to have lost all of its hydrogen but to have retained heavier gases such as nitrogen, oxygen, and water vapor; moreover, their relative proportions were expected to be similar to those found on Earth (78 percent nitrogen, 21 percent oxygen, and 1 percent other gases, including argon and water vapor).

Various attempts were made to determine the surface pressure of the Martian atmosphere, but, unfortunately, all involved indirect measurements. An early calculation, published by Lowell in 1908, was based on the planet's albedo—its ability to reflect light. Lowell estimated what the total reflectivity of the Martian surface would be if there were no atmosphere at all. Since almost five-eights of the planet was desert (albedo 0.16), and three-eights consisted of blue-green areas (albedo 0.07), this came out to be around 0.13, whereas the actual albedo of Mars was 0.27. Correcting for the fact that some of the sunlight would be absorbed by the atmosphere before it reached the ground, Lowell gave a figure of 0.10 for the albedo of the surface. This left 0.17 for the albedo of the atmosphere, which he compared with an estimated albedo of the Earth's atmosphere of 0.75. Because albedo is a measure of the scattering of light, Lowell estimated that the mass of the atmosphere per unit area of Mars would be 0.17/0.75 = 0.23 of the mass per unit area of the terrestrial atmosphere. Given the fact that the force of gravity on Mars is 0.38 times that of Earth, the atmospheric pressure came out to be 0.23 ×

0.38 = 0.087 of the Earth's barometric pressure, or 87 millibars (1 bar is approximately equivalent to the atmospheric pressure of Earth).[7]

In general, later investigators confirmed this result. De Vaucouleurs, for instance, using a method that depended on how the brightness of various patches varied with distance from the central meridian of the planet (since a feature near the limb would be seen through a greater thickness of atmosphere), arrived at a figure right around 85 millibars.[8]

Attempts were also made to determine the composition of the Martian atmosphere. Though nitrogen is notoriously difficult to detect with the spectroscope, oxygen and water vapor should be detectable if they are present. W. W. Campbell's negative results in 1894 at Lick Observatory and in 1909 from the summit of Mount Whitney have already been discussed. The next careful study was made by Walter S. Adams and Theodore Dunham with a spectroscope attached to the 100-inch (2.54-m) reflector at Mount Wilson Observatory in the 1930s. Instead of comparing the Martian spectrum with that of the Moon, as earlier spectroscopists had done, Adams and Dunham used a more reliable method, first proposed by Lowell, which depends on the Doppler effect. The upshot is this: if a body is approaching the Earth, the lines in its spectrum will be shifted slightly toward the violet end of the spectrum; if it is receding, the lines will be shifted slightly toward the red end. In the case of Mars, its orbital motion relative to the Earth produces displacements in the positions of lines in its spectrum compared with those due to the atmosphere of Earth. These displacements are greatest when Mars is near quadrature—that is, at right angles relative to the Earth and Sun. But despite using sensitive instruments, Adams and Dunham failed to identify any lines in the spectrum of Mars that indicated the presence of either oxygen or water vapor.[9]

The first substance positively detected in the Martian atmosphere was neither oxygen nor water vapor but carbon dioxide, and it was found by Gerard Peter Kuiper at the McDonald Observatory in Texas in 1947. By subtracting out the Earth's contribution to the carbon dioxide bands in the spectrum of the Moon, Kuiper found that there was about twice as much carbon dioxide on Mars as on Earth. This discovery was surprising at the time, but it did little to change the overall view of the planet. After all, since carbon dioxide accounts for only 0.03 percent of the Earth's atmosphere, most astronomers naturally assumed that it must be a minor constituent on Mars as well. In 1950, the best guess as to composition of the Martian atmosphere was made by de Vaucouleurs: nitrogen, 98.5 percent; argon, 1.2 percent; carbon dioxide, 0.25 percent; and oxygen, less than 0.1 percent.[10]

THIN AS it was, however, the Martian atmosphere was obviously substantial enough to support clouds. First, there were the yellow clouds, so-called because they appeared bright when observed through a yellow filter. Actually the name is somewhat misleading because almost everything on Mars appears bright in yellow light; it is better to refer to them simply as dust clouds. Some of these clouds may have been seen by Beer and Mädler as long ago as the 1830s, though on the whole this seems rather unlikely, and the first really satisfactory observations were made by Schiaparelli in 1877.

The clouds were particularly well observed in 1909. Fournier noted clouds as early as June, and Antoniadi recorded on August 12 that Mars appeared "citron yellow, and it was nearly impossible to see patches normally as dark as the Mare Tyrrhenum, Syrtis Major and the Sinus Sabaeus! It was truly a unique spectacle. On the following days Mare Cimmerium began to be lightly obscured, and the yellow cloud again covered almost the whole of the planet. . . . Only the Mare Sirenum retained its normal intensity."[11] At the following opposition, in 1911, Antoniadi found large parts of Mare Erythraeum covered for several weeks with yellowish clouds, and again in August 1924 the planet was almost completely covered "and presented a cream color similar to that of Jupiter."[12] In Antoniadi's drawing made on December 24, the dark patch Nodus Gordii—now known to be one of the Martian shield volcanoes— was virtually the only visible feature on the disk.

Antoniadi thought that the yellow clouds consisted of fine dust carried aloft by the winds. They were, he noted, most often seen when Mars was near perihelion. This was only to be expected; after all, at perihelion the solar heat was half again as great as at aphelion, and this ought to produce stronger winds, so that the lighter sand and dust particles would be more easily picked up from the Martian deserts.[13]

There were major dust storms in 1941 and 1956 as well. The latter began on August 20 with a bright cloud over the Hellas-Noachis region that was discovered at both the Kwasan Observatory in Kyoto and the National Science Museum in Tokyo, Japan. This cloud quickly expanded through the southern hemisphere, and by mid-September, when Mars came to opposition, had developed into a planet-encircling storm—observers at the time noted, as Antoniadi had in 1924, that the disk had become virtually blank.

Apart from the dust clouds and planet-encircling storms, observers have also noted whitish clouds (also referred to as "bluish" clouds, because they appear most brilliant when the planet is observed through a blue filter). Every autumn and winter a polar hood forms over the north polar cap, often extending as far south as latitude 50° N, at which time it obscures much of Mare Acidalium. In addition, limb hazes are commonly seen in spring and summer at the planet's morning limb. Still other whitish clouds form predominantly in the late afternoon, disperse during the night, and then re-form the following afternoon; since they frequently appear over the same region for days at a time, these are referred to as "recurrent" clouds. Often they resemble a light mist; at other times they become so bright that they rival the polar caps. Areas commonly affected by recurrent clouds include Syrtis Major, Elysium, Arcadia, Nix Olympica, and the Tharsis region. The latter is the site of the famous "W" clouds (as seen in an inverted telescopic image; right-side up they would be "M" clouds), first photographed by E. C. Slipher at the 1907 opposition. This group of clouds has been observed many times over the years, although the exact W (or M) pattern has seldom been repeated. By analogy to the Earth, observers assumed that these recurring clouds formed preferentially over elevated areas —that is, that they were orographic clouds made of water-ice crystals which formed as moist air rose up windward slopes.

CURIOUSLY, the Martian atmosphere did not seem to be equally transparent in all wavelengths. When photographed through filters, the familiar surface details of Mars always showed up well in red and yellow, but (with the exception of the polar caps) they usually disappeared in blue, violet, and ultraviolet. This was noted as long ago as 1909 by C. O. Lampland at the Lowell Observatory, and it was long thought to imply that there was something in the Martian atmosphere that acted as a screen to block out short-wavelength radiation, in the same way that stratospheric ozone does on Earth—

though, given the scarcity of atmospheric oxygen on Mars, no one ever seriously suggested that ozone itself was responsible. The supposed screen was referred to as the "Blue Haze" or the "Violet Layer," though actually the name was something of a misnomer because there was no visible haze or layer. The importance of such a screen to the question of life on the planet was obvious: if the haze existed, it would provide some protection against damaging ultraviolet radiation from the Sun.

Beginning in 1926, E. C. Slipher began to suspect that the Violet Layer was not permanent; at that opposition and again in 1928, the dark areas were revealed at times in blue photographs. However, "extreme caution suggested waiting for more tell-tale examples . . . to thoroughly clinch the evidence."[14] The telltale example came in May 1937, when, for several days around opposition, the dark markings, including Syrtis Major and Sinus Sabaeus, were as conspicuous in photographs taken with a blue filter as in those taken with yellow or red filters. Later studies showed other blue clearings; sometimes they seemed to affect only one hemisphere of Mars, at other times the whole planet. Oddly enough, many of the more remarkable episodes seemed to take place at or close to the date of the opposition. After reviewing several thousand photographs taken over a number of years, Slipher concluded that "the clearings were so striking in character and were so closely associated with the oppositions that such occurrences were accepted to be the rule."[15]

At that time the dark areas were generally believed to be tracts of vegetation. In 1941, Seymour L. Hess pointed out that if the Violet Layer actually screened out shortwave radiation, then the growth of vegetation ought to be arrested during a blue clearing. In October of that year, Hess found that this actually seemed to be the case; during a blue clearing, the wave of darkening seemed to halt for several days, and resumed only after conditions had again returned to normal—a development that seemed to furnish very strong arguments both for the existence of a Violet Layer and for the presence of vegetation on Mars.

What was the Violet Layer? There were various theories, including one favored by Kuiper and Hess that it consisted of tiny ice crystals. The Estonian-born astronomer E. J. Öpik suggested that it might be made up of a mixture of surface dust and small particles of carbon black formed by decomposition of carbon dioxide by solar ultraviolet radiation. The point is that no one really knew, and the whole question remained among the unsolved mysteries of Mars until after the beginning of the spacecraft era.

We now know, however, that the Violet Layer does not exist. The blue clearings have been rather mundanely explained as due to phase-angle effects of light-scattering by airborne dust, which causes occasional enhancement of the low-contrast differences between the light and dark areas in blue light.[16]

THOUGH WATER vapor in the Martian atmosphere eluded detection by spectroscope, there was never much doubt that water did exist on Mars. The Ranyard-Stoney theory, that the polar caps were frozen carbon dioxide, had never really caught on; most investigators still supported William Herschel's idea from 1784 that they were water ice. Antoniadi went so far as to state, "This theory of Herschel's is correct."

Lowell had made much of a dark band which he observed accompanying the cap in its retreat. "At pressures of anything like one [Earth] atmosphere or less," he wrote,

carbon dioxide passes at once from the solid to the gaseous state. Water, on the other hand, lingers on in the intermediate stage of a liquid. Now, as the Martian cap melts it shows surrounded by a deep blue band which accompanies it in its retreat, shrinking to keep pace with the shrinking of the cap. . . . This badge of blue ribbon about the melting cap, therefore, shows conclusively that carbon dioxide is not what we see, and leaves us with the only alternative that we know of: water.[17]

Antoniadi was skeptical of the reality of Lowell's band, which he regarded as a mere contrast effect. But de Vaucouleurs, from his observations in 1939, found that the band was not generally visible in winter, when the cap appeared largest; it presented its full development only during the spring, which could hardly have been the case were it illusory.[18] In April 1950 Kuiper came to the same conclusion after observations with the 82-inch (2.08-m) refractor at the McDonald Observatory.

That a band of some sort existed was, however, hardly definite proof that it consisted of water. The blue color alleged by Lowell was a particularly weak argument in support of this conclusion, as Alfred Russel Wallace had pointed out in *Is Mars Habitable?* in 1907: "It is perfectly well known that although water, in large masses and by transmitted light, is of a blue colour, yet shallow water by reflected light is not so; and in the case of the liquid produced by the snow-caps of Mars, which the whole conditions of the planet show must be shallow, and also be more or less turbid, it cannot possibly be the cause of the 'deep blue' tint said to result from the melting of the snow."[19]

But though inferences based on the color of the Lowell band did not hold water, so to speak, most investigators continued to hold firmly to the notion that the caps were water ice. Kuiper in 1948 showed that their spectrum resembled that of ordinary snow, and not carbon dioxide snow, which was quite different. Thus he concluded: "The Martian polar caps are not composed of carbon dioxide and are certainly composed of water frost at low temperature [much below 0°C]."[20] Moreover, the alternative—caps of frozen carbon dioxide—seemed even more unlikely because it did not seem possible that the temperature of Mars, even at the poles, ever got cold enough for carbon dioxide to freeze. Although various attempts had been made to calculate the temperature of Mars on theoretical grounds—including Lowell's well-known "warm as the south of England" estimate of 1907—the first direct measures of the temperature of the planet date only from the 1920s. For this purpose D. H. Menzel, W. W. Coblentz, and C. O. Lampland at the Lowell Observatory and E. Petit and S. B. Nicholson at Mount Wilson used a thermocouple, an instrument that consists of a circuit formed with wires of different metals soldered end to end. If a very minute amount of heat—say, from part of a planet's surface—warms one of the two joints, a tiny electric current is set up that can be measured with a galvanometer, and this leads to a determination of the planet's surface temperature. The thermocouple results showed that the temperature in the bright areas at the equator of Mars sometimes reached 10–20°C (50–68°F), and reached 20–30°C (68–86°F) in the dark areas, while the temperature in the south polar region in 1924 was –50–70°C (–58–94°F). At the low atmospheric pressures on Mars, a temperature of at least –100°C (–148°F) would be required for solid carbon dioxide to form a deposit on the surface. Thus the conclusion once again seemed to follow: the caps must be water ice.

EVEN IF they were formed of water ice, however, the caps undoubtedly contained only minute quantities of water. From the rapidity of their melting, de Vaucouleurs estimated that they were only a few centimeters thick. Thus, the planet was obviously extremely arid.

Lowell had compared the salmon-colored areas of the planet to the Painted Desert of Arizona. There had never been any doubt that this is what they were—dusty deserts, desolate almost beyond the ability of anyone to imagine and vast by terrestrial standards, since altogether they occupied an area ten times larger than the combined areas of the Saharan, Libyan, and Nubian Deserts on Earth. As for their composition, the pioneering polariscopic observations of Bernard Lyot, who studied the specific way light was reflected from Mars and compared it with that of various minerals, indicated that the deserts might be covered with a layer of volcanic ash. Later, Audouin Dollfus demonstrated a close match with pulverized limonite, a compound rich in iron oxides.

The seasonal wave of darkening affecting the dark areas also seemed to be well established. Again, Lowell's observations were largely confirmed, and indeed embellished, by a number of other skillful observers of the planet. Thus Fournier wrote: "As the spring advances, the dark shading progressively encroaches from the pole towards the equator across the seas. This advance, across wide expanses and along channels, takes place at a variable speed, but nearly always very rapidly; a few weeks only suffice to change the landscape completely."[21]

Some of the markings seemed to change in a regular way with the seasons—for example, Antoniadi claimed that Syrtis Major appeared broad and dark in winter, narrow and weak in summer; Pandorae Fretum underwent a pronounced darkening each summer, and so on. Still other changes did not appear to be seasonal. Thus Nepenthes-Thoth showed a dramatic broadening in 1888 and again between 1911 and 1929; and Solis Lacus, whose classical form had been an ellipse with its long axis lying east-west, underwent a marked transformation in 1926 so that its long axis ran north and south. These changes were carefully recorded by Antoniadi, who was certain that they were due to "an invasion by the dark or greenish vegetation on to the lightly-shaded or rosy areas adjacent to the great dark areas."[22]

The changes of color that affected the dark markings also seemed to be seasonal. Lowell, to whom the disk presented "washes of color, the one robin's-egg blue, the other roseate ochre," had remarked that some of the dark areas changed from bluish green to brown and then to yellow with the arrival of midsummer in the southern hemisphere, reminding him of the changes observed in terrestrial vegetation during dry summers and autumn. Antoniadi paid no less minute attention to the coloring of the planet. In 1924, he and his associate Fernand Baldet described complex color changes spreading from the polar regions. "Not only the green areas," Antoniadi wrote, "but also the greyish or blue surfaces, turned under my eyes to brown, lilac-brown or even carmine, while other green or bluish regions remained unaffected. . . . It was almost exactly the color of leaves which fall from trees in summer and autumn in our latitudes."[23]

These observations were somewhat suspect because they were made with large refractors, in which chromatic aberration can never be entirely eliminated. Observations with reflectors were less objectionable. Thus E. E. Barnard, using the 60-inch (1.52-m) reflector at Mount Wilson in 1911, described the disk of Mars as "feeble salmon," while dark markings such as Syrtis Major and Mare Tyrrhenum were "light-gray."[24] G. P. Kuiper, using the 82-inch reflector at McDonald Observatory, similarly found that in the spring of 1956 the dark areas were predominantly neutral gray, although he also noted "a touch of moss green" in some of the equatorial regions.[25]

The greenish colors naturally suggested chlorophyll. Chlorophyll looks green because it reflects light in the green part of the spectrum; however, it absorbs strongly in the infrared. Therefore, if the dark areas on Mars contained chlorophyll-bearing plants, they ought to look dark in infrared photographs. In fact, however, just the opposite was true. Thus, any plants on Mars could not contain chlorophyll; instead, they were widely believed to be primitive but hardy organisms, perhaps similar to terrestrial lichens.

Various attempts were made to detect the presence of organic materials spectroscopically, and there was a brief flurry of excitement in 1959 when W. M. Sinton announced that he had found strong absorption bands in the spectrum of the dark areas that seemed to coincide with those produced by carbon-hydrogen bonds in organic molecules. Alas, the Sinton absorption bands were later found to be caused by molecules in the terrestrial atmosphere; they had nothing at all to do with Mars.[26]

The vivid colors, too, finally proved to be illusory. The true tones of the Martian dark areas are drab, dull reds and gray-browns; the greens and blues are mere results of a visual physiology effect known as simultaneous contrast, which causes a relatively neutral-toned area surrounded by a yellow-orange field to appear blu-ish green to the eye. The effect has been well known (except to astronomers, evidently!) since the nineteenth century—it was first pointed out by a French chemist named M. E. Chevreul in 1839.[27] Chevreul, the director of dyeing at France's national tapestry work-shop (Manufactures Royales des Gobelins), had been charged with improving the intensity and fastness of dyes, and realized that the apparent brightness of a color depends more on the colors sur-rounding it than on the intensity of the color itself. Thus, he wrote, "where the eye sees at the same time two contiguous colors, they will appear as dissimilar as possible, both in their optical compo-sition and in the height of their tone. We have, then, at the same time, simultaneous contrast of color properly so called and con-trast of tone."[28] Simultaneous contrast was briefly considered by nineteenth-century astronomers John Herschel and François Arago as an explanation of the greenish blue colors of Mars, but then promptly forgotten. These subjective colors, together with the line-like visions of the color-blind Italian astronomer G. V. Schiaparelli, created the compelling dream of a living world that so long domi-nated our views of Mars.

ALTERNATIVE (that is, nonvegetative) explanations of the in-tensity changes in Martian surface markings began with the view of Svante Arrhenius, the Nobel Prize–winning chemist, who sug-gested in 1912 that certain minerals known as hygroscopic salts might be responsible; these salts absorb water and show dra-matic darkening on contact with it. But Arrhenius's idea never received much support. Another theory was worked out by Uni-versity of Michigan astronomer Dean B. McLaughlin. In a series of papers written between 1954 and 1956, McLaughlin proposed that Mars was actively volcanic. According to this theory, dark ash that spewed from volcanic vents flared downwind, in the process giving rise to the characteristic caret-shaped dark areas on Mars such as Sinus Meridiani (Dawes' forked bay). Seasonal variations in wind direction produced a redistribution in the primary ash deposits, which changed the shape and darkness of the Martian markings. Thus, according to McLaughlin, the well-documented changes were due to windblown dust rather than to the annexation of deserts by vegetation.[29]

After McLaughlin put forward his theory, Tsuneo Saheki, a Japanese astronomer, went so far as to suggest that a very bright "flare" that he had recorded at the Martian terminator on December 8, 1951, had been nothing less than a Martian volcano caught during an actual eruption. Though the volcanic explanation is doubtful at best, Saheki's description is nevertheless interesting: "When I first looked at Mars, . . . I saw Tithonius Lacus just inside the limb. Very soon afterwards, a very small and extremely brilliant spot became visible at the east end of this marking. At first I could not believe my eyes, because the appearance was so completely unexpected."[30] The spot brightened further until, briefly, it surpassed the north polar cap itself; however, in less than an hour it had faded completely from view.[31]

G. P. Kuiper roundly criticized McLaughlin's theory, in large part because the existence of active Martian volcanoes seemed to be incompatible with the extreme scarcity of water vapor in the planet's atmosphere.[32] It still seems so today, but at least McLaughlin had the kernel of an important idea. Kuiper himself, who had hitherto embraced the vegetation hypothesis, changed his mind within the year, in large part because of the impression made on him by the great dust storm of 1956. The obvious ability of winds to move dust on Mars led him to propose that the dark areas might be dust-covered lava fields. As McLaughlin had in his original theory, Kuiper invoked seasonal removal of the dust by wind currents to explain the "wave of darkening."[33] In 1958, a Russian astronomer named V. V. Sharanov arrived independently at the same explanation. The air currents on Mars, he wrote, "vary from season to season, depositing dust at some times of the year and blowing it away at other times. Thus, for instance, the inherently dark surface . . . may brighten at a definite time of the year as a result of settling of light-colored dust blown over from the desert areas."[34]

JUST AS the idea that only the vegetation hypothesis could account for the observed phenomena of Mars was beginning to be called into question, other basic tenets of the Lowellian Mars were also falling by the wayside. Lowell had argued that the dark areas were dry sea bottoms, but radar studies in the early 1960s indicated that at least some of them were elevated rather than low-lying areas. This, incidentally, destroyed yet another argument once regarded as compelling support of the vegetation theory. In 1950, E. J. Öpik had pointed out that if the dark areas were low, they ought to be quickly covered by yellowish dust unless they had some means of regenerating themselves. At the time this was thought to prove that they were tracts of vegetation. If, however, the dark expanses were actually elevated areas, there was no need to believe this—the dust might just as well be removed by wind scouring.

The long-standing belief in a flat Mars, based on the smoothness of its terminator as observed from Earth, was also called into question. Clyde Tombaugh, a former assistant astronomer at Lowell Observatory and later a professor at New Mexico State University, argued in 1961 that mountains on Mars would not be detectable from the shadows they cast at the terminator unless they were at least 27,500 ft (8,500 meters) tall, and more rounded features would have to be even taller. Incidentally, Tombaugh, Öpik, and Ralph B. Baldwin all independently suggested around 1950 that there might be numerous impact craters on Mars.[35]

The century-long quest for Martian water vapor finally came to fruition at the mid-winter opposition of 1963. The detection of very minute amounts of water vapor in the Martian atmosphere was announced by Audouin Dollfus, who set up a special telescope at the Jungfraujoch high in the Swiss Alps, and by H. Spinrad, G. Münch, and L. D. Kaplan, who used a photographic emulsion especially sensitive to infrared radiation to record the spectrum of Mars near quadrature with the 100-inch (2.45 m) reflector at Mount Wilson.[36] The latter team found that the average amount of precipitable water on Mars (that is, the equivalent thickness of liquid water if all the atmospheric water were condensed onto the surface) was only about 14 micrometers, compared with 1,000 micrometers of precipitable water in even the driest desert areas of Earth. In addition, they estimated that the partial pressure of carbon dioxide on Mars was 4.2 millibars and that the total atmospheric pressure at the surface could not be more than about 25 millibars.[37] This was a drastic downward revision from the previously accepted figure of 85 millibars. Thus, Mars was drier and had an even thinner atmosphere than anyone had realized.

Parenthetically, the reason so many of the earlier investigators were so far wrong about the thickness of the Martian atmosphere was that most based their estimates on albedo studies. They had assumed the Martian atmosphere to be usually clear and transparent, but this is not the case; often there is at least some dust present, which makes the atmosphere more reflective than they had supposed.

By the time Spinrad, Münch, and Kaplan published the results of their monumental study, a very different view of Mars was beginning to come into focus. In many ways this new Mars resembled the modern view of the planet: a bone-dry world with an extremely rarefied atmosphere, a surface with perhaps considerable relief, and changes in its markings that were the result of windblown dust rather than vegetation. However, the paradigm shift was not yet complete by 1965 when the first spacecraft reconnaissance of Mars took place, and most astronomers and the lay public were shocked, even depressed, by what it revealed.

Spacecraft to Mars

The last of the oppositions of Mars that took place before spacecraft visited the planet and forever destroyed our innocence occurred in March 1965. It was an aphelic opposition, and it was also the first that I personally observed with a telescope; I was then a ten-year-old equipped with a modest 2-inch (5-cm) refractor. The instrument was too small to show much, but I was excited to be able make out a few of the dark patches, which at the time were still generally believed to be tracts of vegetation. Moreover, at that moment there was a mood of great expectation—two spacecraft were then headed toward Mars: the American *Mariner 4* and the Russian *Zond 2*.

THE SPACE AGE officially began with the launch of the first satellite, *Sputnik 1,* by the Russians on October 4, 1957. The first successful probes to the Moon were launched in 1959, followed by rockets to Mars and Venus. The Russians had the clear edge in rocket development, and they attempted to launch a rocket toward Mars as early as October 1960; unfortunately, it failed to reach Earth orbit. Since the Russians made a point of shrouding their space program in secrecy, they published no information about it at the time.

The Russians launched a Venus probe in February 1961 but lost contact with it long before it reached its destination. They followed up with two more Mars probes in the fall of 1962. The first failed to reach Earth orbit, and again, nothing more was said about it; but the other, *Mars 1,* a spacecraft weighing 874 kilograms and equipped with television cameras, was placed in the correct orbit, and at first everything seemed to be going well. (Since we are now in the modern era, I propose henceforth to drop imperial equivalents and give values in metric units only.)

The basic principles involved in launching an interplanetary probe such as *Mars 1* are straightforward enough. Note that when an object is launched from the Earth's surface at relatively slow speeds, it simply follows a curved trajectory back to the Earth. At higher and higher speeds the curvature of the arc becomes more and more gentle, until there comes a certain point—at a velocity of 28,000 kilometers per hour, to be exact—when the rate at which the rocket is traveling forward and dropping downward equals that at which the Earth's surface below curves away from it. Thus, though it continues to fall freely, the rocket never reaches the ground (for the moment, we are ignoring air friction). At this point it has achieved Earth orbit. If the rocket is further accelerated to a speed of 11 kilometers per second, or 40,000 kilometers per hour—the escape velocity—it escapes from the sphere of the Earth's gravitational influence and becomes an independent body traveling in its own orbit around the Sun.

The fuel supply of a rocket is very limited, however, and the only practical way for it, or any spacecraft, to reach another planet is for it to be placed in what is known as a transfer orbit, so that it can coast, without using any fuel, for most of the journey. If it is to travel to the inner planets, Mercury and Venus, the spacecraft must be slowed down slightly relative to the Earth; in order to reach the outer planets, such as Mars, it must be speeded up.

The most energy-efficient trajectory between two planets is the so-called Hohmann transfer ellipse, named after the German engineer W. Hohmann, who first described it in 1925. If the orbits of the Earth and Mars were exactly circular, the Hohmann transfer ellipse would be a path in which the spacecraft left the Earth at an angle tangential to its orbit and arrived at an angle tangential to the orbit of Mars. This orbit would have its perihelion at the launch point (Earth) and its aphelion at the orbit of Mars; the spacecraft's period of revolution around the Sun would be 520 days, and in getting from the Earth to Mars it would travel halfway around this ellipse, so that the transit time from the Earth to Mars would be 260 days. Since in this time Mars would have moved a distance of (260/687) ×

$360° = 136°$ along its orbit around the Sun, it follows that in order for the spacecraft to reach Mars, the relative positions at launch must be such that the Earth-Sun-Mars angle is $180° - 136° = 44°$.

Such conditions occur about fifty days before each opposition, and they define the "launch window." Of course, the actual orbits of the Earth and Mars are not exactly circular, nor do the two planets lie in quite the same plane; thus, the actual conditions will vary somewhat from launch window to launch window—in particular, less energy is required to reach Mars during launch windows at which a perihelic opposition occurs. Also, since the minimum energy requirement for the spacecraft to reach Mars is actually rather modest, it is possible for the trajectory to depart considerably from the Hohmann transfer ellipse. Without going into details, it is fair to say that in general, launch windows occur roughly two or three months before opposition. Thus, since there was to be an opposition in January 1963, the Russians were within the launch window in getting *Mars 1* under way at the beginning of November 1962.

Mars 1 remained in contact with the Earth until March 21, 1963, by which time it reached a distance of 106 million kilometers away in space. Unfortunately, radio communications were then suddenly lost, and the following June the spacecraft passed silently and uselessly by the red planet at a distance of 195,000 kilometers; no photographs or other data were obtained.

The next launch window occurred in the fall of 1964. By then, the Americans had also become active. They had already sent the first successful interplanetary probe, *Mariner 2*, past Venus in December 1962, and it had made measurements showing the high surface temperatures (up to $477°C$) and generally sinister conditions that prevail on that planet. In November 1964, the United States prepared to launch two Mars probes. The first, *Mariner 3*, set out on November 5, but the fiberglass shroud designed to protect it during its ascent through the Earth's atmosphere failed to eject, and the extra weight prevented the probe from achieving the proper transfer orbit. It was also unable to deploy its solar panels and soon ran out of power. The backup spacecraft, *Mariner 4*, was launched on November 28—the 305th anniversary, incidentally, of the day Christiaan Huygens drew a rough sketch of the Syrtis Major region. This time everything went well; the shroud was ejected, and the solar panels deployed. Two days later, the Russians followed with the launch of their latest Mars probe, *Zond 2*.

Mariner 4 led the way to Mars and was due to arrive there the following summer, some three weeks ahead of its Russian counterpart. But the Russians had not yet solved their problems with communications, and in early May 1965 their experience with *Mars 1* was repeated—they lost contact with *Zond 2*, and it was never heard from again.

Fortunately, *Mariner 4* continued to function as planned, and on July 14, 1965, made a close sweep past Mars. The American engineers had kept the design of the spacecraft simple; it weighed only 260 kilograms and carried a television camera and other scientific instruments, including a magnetometer and trapped-radiation detector (to measure the intensity of any magnetic fields and radiation belts around Mars). The first of twenty-two images was obtained when the spacecraft was 16,900 kilometers from the surface—it was a view of the limb, showing a section of the Amazonis desert near the dark patch Trivium Charontis. The rest of the series covered a discontinuous swath extending south and then eastward from Amazonis across Zephyria, Atlantis (the brightish region between the dark areas Cimmerium and Sirenum), Phaetontis, and Memnonia. The last three pictures were taken beyond the terminator of Mars from a distance of 11,900 kilometers. Although the images were of low contrast and rather murky—possibly because of a light leak in the camera system—they were good enough to show a distinctly cratered surface, of which the largest crater, in Mare Sirenum, measured 120 kilometers across. In all, about three hundred craters were recorded—but, I hasten to add, nary a canal!

In every way the *Mariner 4* results came as a shock. The probe seemed to reveal a Mars that was, in a word, moonlike. The surface appeared old and dead, and apparently had not changed appreciably for billions of years. This dour impression was reinforced by the results of the S-band radio occultation experiment. Two hours after it made its closest approach to the surface of Mars (9,850 km), *Mariner 4* passed behind the planet, at a point on the sunlit side between Electris and Mare Chronium. Its radio signal was distorted by its passage through the thin Martian atmosphere, and this was repeated when it emerged again from the night side, at a point above Mare Acidalium. By analyzing the shape of this distortion it was possible to calculate the surface pressure at the two occultation points. The result was distressingly low—only 4.0 to 6.1 millibars.[1] When these data were combined with the earlier ground-based work of Kaplan, Münch, and Spinrad, who, as we have seen, showed the partial pressure of carbon dioxide on the surface of Mars to be around 4 millibars, it became clear that the Martian atmosphere must be made up of something like 95 percent carbon dioxide—thus it seemed only too likely that Ranyard and Stoney had been right after all in surmising that the polar caps must consist of frozen carbon dioxide instead of ordinary water ice.[2] At that low pressure, liquid water, even in the relatively warm equatorial regions of the planet, would not be stable on the surface. In their official summary of the results of the *Mariner 4* mission, R. B. Leighton and the other members of the television team argued that "the heavily cratered surface of Mars must be very ancient—perhaps 2 to 5 billion years old . . . [and] it is difficult to believe that free water in quantities sufficient to form streams or fill oceans could have existed on Mars since that time."[3]

IN RETROSPECT, it is somewhat surprising that the cratered surface of Mars evoked such great surprise. After all, impact is now known to be the dominant force in fashioning the early surfaces of the Moon and planets, and we have actually witnessed comet fragments crashing into Jupiter. In the early 1960s, however, the details of all this were still being worked out. Even the best known impact feature on Earth, the kilometer-wide crater near Winslow, Arizona, was not definitively shown to be the result of an impact until the late 1950s.[4] The origin of the lunar craters was still hotly debated, and it was by no means certain that they had been formed by impact rather than by what can be described, broadly speaking, as volcanic processes. After *Mariner 4* revealed the Martian craters, the debate extended to that planet as well. Only with the Apollo spacecraft missions was the issue finally resolved in the case of the Moon, and we can now say for certain that the craters, on both the Moon and Mars, were formed by impacts.[5]

As I mentioned earlier, a few farsighted astronomers such as Clyde Tombaugh, E. J. Öpik, and Ralph Baldwin had actually predicted the existence of craters on Mars. One, John E. Mellish, went so far as to claim that he had directly observed the Martian craters in November 1915, when the disk was only 7.8″. He had used the 40-inch (1.02-m) Yerkes refractor just after sunrise with a magnifying power of 1,100×. Mellish did not publish an account of these observations until the year after the *Mariner 4* flyby, though he had written earlier about them to other astronomers.[6] He also implied that his onetime colleague E. E. Barnard had seen the Martian craters while he was at the Lick Observatory in the 1890s. Unfortunately, Mellish's drawings were destroyed in a fire; Barnard's, however, turned up at Yerkes a few years ago (they show some dark circular patches ["oases"], but no true craters).

Mellish's observation sparked controversy. Some astronomers were convinced that he had seen actual craters on Mars; others were frankly skeptical.[7] Obviously he saw something; presumably it was some of the bewildering details—small masses and dark spots—into which the Martian surface resolves under the best viewing conditions. Never having seen anything like it before, he was understandably astonished and found intimations of strikingly unfamiliar aspect. But his specific claim to have seen craters can now be dismissed. It is quite impossible to observe any topographical relief directly from Earth; this the Hubble Space Telescope observations have now put beyond doubt. The presence of the Martian atmosphere smooths out the evidence of surface irregularities at the terminator, and moreover, though Mars does indeed have mountains, the tallest of them—including towering Olympus Mons, the tallest mountain in the solar system—are great volcanic shields rather than peaks; their slopes are very gradual, and they cast no visible shadows even at the terminator.

THE *Mariner 4* results spurred astronomers to completely revise their ideas about Mars; it was now obvious that earlier views had been very wide of the mark indeed. The vegetation theory was dead —the Martian environment seemed to be too hostile to support vegetation, and even the existence of the wave of darkening, which "took second place only to the Martian canals in historical development of the life on Mars hypothesis,"[8] was doubtful. Though there could be no question that there were considerable changes in the form and intensity of the dark areas, the existence of an actual "wave" extending from the south pole to the equator each spring seemed at best only weakly supported by the observations. A study by Charles F. Capen of the Lowell Observatory showed that the greatest intensity changes actually seemed to occur in the light areas, and only the dark areas in the immediate vicinity of the south polar cap showed the classical pattern of appearing consistently lighter in summer than in spring. Systematic latitudinal changes in the intensity of features with the seasons, at least, did not seem to occur, and any changes that did take place were evidently due to windblown dust rather than vegetation.[9]

The changing perspective of Mars brought about by the *Mariner 4* results is well illustrated by the before and after views of Capen, who was one of the most skillful Mars observers of that time. In 1964–65 he gave a description of Mars that was in many ways "classical." Thus he plotted numerous canals and noted vivid color changes; for instance: "The Syrtis Major was changing from a blue-green to a green-blue hue. . . . The Mare Acidalium changed from its winter shades of variegated gray and brown to its spring coloration of dark gray and blue-gray shades with gray-green oases."[10] In May 1969, with Mars near opposition, Capen enjoyed a series of splendid views with the 82-inch (2.08-m) reflector at McDonald Observatory in Texas. "When Mars was first brought into focus on the nights of May 29, 30, and 31, its globe appeared to be draped in a dark gray spiderweb, resembling the shade and texture of iron-filings. When 1,000× was employed the global network . . . resolved into dark circular features and parallel aligned streaks, some of which were fortuitously aligned into canal-like lineaments." These details he afterward identified with large craters and composites of dark blotches and streaks shown on the spacecraft photographs.[11]

AFTER *Mariner 4*, the next probes sent to Mars were *Mariners 6* and *7* in 1969 (*Mariner 5* had gone to Venus). *Mariner 6* set out on February 24 and *Mariner 7* on March 27. We now know that there were two unsuccessful Russian launches that spring as well. They were not even announced at the time, and there is no reason to say anything more about them.

The two Mariner spacecraft, like *Mariner 4* before them, were designed as flyby missions, but they began imaging Mars well before their arrival. These far-encounter views had a resolution somewhat better than the best Earth-based images but were still too blurry to show the Martian surface features very well—for instance, Nix Olympica, which had first been seen by Schiaparelli in 1879 as a tiny white spot, appeared as a large, bright ring, interpreted at the time as the outline of a large impact crater.

The actual Mars encounters took place only a few days after the *Apollo 11* flight, during which Neil Armstrong and Edwin Aldrin became the first men to walk upon the surface of the Moon. *Mariner 6* arrived first, on July 31, 1969, and provided twenty-five close-up photographs of the equatorial region between longitudes 60° w and 320° w, including Aurorae Sinus, Pyrrhae Regio, and Deucalionis Regio. *Mariner 7* followed on August 5 and obtained thirty-three photographs of a region along the edge of the south polar cap and another swath of the planet—again mostly in the southern hemisphere—covering Thymiamata, Deucalionis Regio (again), Hellespontus, Hellas, and Mare Hadriaticum. In all, the two probes increased the close-up coverage of the planet from *Mariner 4*'s 1 percent to about 10 percent, and their photographs were much clearer than those of *Mariner 4*—mainly because of improved techniques, but partly because the spacecraft passed closer to the planet, within less than 3,500 kilometers.

The by now familiar craters were fairly ubiquitous in the photographs, and included a number of frost-covered features along the edge of the polar cap. For the most part, the moonlike Mars of *Mariner 4* was very much in evidence, but there were also some surprises. One interesting area was Hellas, a large, circular bright area some 1,300 kilometers across. Classically it had been regarded as an elevated area, since it was a site where whitish clouds frequently developed, and it had often appeared to be frost covered in winter; however, we now know that it is a low-lying basin, and the Mariner photographs showed it to be surprisingly smooth. Another interesting area, located at 40° w, 15° s, consisted of jumbled ridges, or "chaotic terrain," which appeared to have formed by the withdrawal of subsurface material and collapse of the overlying sediments and rocks.

Apart from the television experiments, other instruments on board the two spacecraft confirmed that Mars's atmospheric pressures are very low—*Mariner 6* measured 6.5 millibars in the Sinus Meridiani region, and *Mariner 7* found only 3.5 millibars over Hellespontica Depressio, a region that had long been regarded as a depression, though in fact it proved to be elevated. The temperature at the south polar cap was measured at $-123\,^{\circ}\text{C}$ ($-190\,^{\circ}\text{F}$), almost exactly what was expected for carbon dioxide ice. At the time most scientists believed that both of the caps consisted entirely of frozen carbon dioxide. Finally, the probes found no trace of a magnetic field.

All in all, the view from the flyby Mariners of 1969 was most discouraging. They had covered only a tenth of the Martian surface, but this had included several of the main dark areas of the planet—and it had been on the dark areas, after all, that observers had always focused their attention, ever since Christiaan Huygens had first sketched Syrtis Major in 1659. We now know that the dark areas of the southern hemisphere contain the most heavily cratered terrain on Mars. And although the cameras of the flyby Mariners took numerous photographs of craters, by sheer chance they missed all of the truly spectacular features of the Martian landscape—the volcanoes, canyons, and dry riverbeds.

As the 1960s ended, the drab and moonlike Mars seemed to have been confirmed by three flyby spacecraft, and the fascinating Lowellian world of dry sea bottoms, canals, lonely deserts, and dying civilizations had faded like a dream. In many ways 1969 was the nadir of Martian studies. But the moonlike Mars was to prove as much an illusion as the Lowellian Mars had been. This brings us to *Mariner 9* and the next great year of Martian discovery—1971.

Mariner 9

As it turned out, the belief in the drab and moonlike Mars that followed in the wake of the flyby Mariners was premature, and quite as unjustified as the belief in the Earthlike Mars that had preceded it. The close-up photographs taken by *Mariners 4, 6,* and *7* had covered only 10 percent of the Martian surface and had somehow managed to miss the most exciting features. In 1968, Clark R. Chapman, James Pollack, and Carl Sagan warned: "If substantial aqueous-erosion features—such as river valleys—were produced during earlier epochs on Mars, we should not expect any trace of them to be visible on the Mariner IV photographs unless they were of greater extent than typical features on the Earth. . . . [A]ny conclusions . . . that the apparent absence of clear signs of aqueous erosion excludes running water during the entire history of Mars . . . must certainly be regarded as fallacious."[1] Their comments could hardly have been more prophetic, although they made little impression at the time.

The next phase in the spacecraft exploration of Mars began in 1971, when five spacecraft—three Russian and two American—were readied for launch. American plans called for putting two spacecraft into closed orbits around Mars. The first, *Mariner 8,* was to enter a highly inclined orbit that would allow coverage of about 70 percent of the surface, including the polar areas. The emphasis was to be on mapping topographic rather than albedo features, thus requiring photography under conditions of low Sun elevation (when shadows would bring out relief). The second spacecraft, *Mariner 9,* would be placed in a more nearly equatorial orbit, from which it could carry out a careful study of albedo variations, best seen under conditions of high Sun.

On May 8, 1971, *Mariner 8* lifted off from Cape Canaveral (actually Cape Kennedy, as it was called for a brief time; the name was changed back to Canaveral a few years later). The second stage of the Atlas-Centaur launch vehicle failed to ignite, however, and the spacecraft crashed ignominiously into the Atlantic Ocean. At once, the mission engineers made changes in their plans for the second spacecraft, *Mariner 9,* in order to try to fulfill with one spacecraft as many as possible of the objectives originally designed for two. The revised plans called for *Mariner 9* to enter an orbit inclined 65° to the Martian equator, with a minimum altitude of 1,350 kilometers. This meant both a greater altitude and higher Sun conditions than were ideal for the topographic mapping, while the elevation of the Sun would not really be high enough for the albedo variations study. All in all, however, it was a reasonable compromise. *Mariner 9* was launched on May 30, 1971, and successfully placed in its transfer orbit for Mars. This spacecraft would ultimately revolutionize all our ideas about the planet.

In the meantime, two days after *Mariner 8* had aborted its mission, the Russians launched their first spacecraft from Baikonur, in central Kazakhstan, but it failed to leave Earth orbit owing to "a most gross and unforgivable mistake" in a command sent to the on-board computer. This was to be an orbiter-only mission, which the Russian space planners had hoped to send on a faster trajectory to Mars so that it would arrive ahead of the American Mariners. The Russians followed up with two more spacecraft, both orbiter-landers. After entering Martian orbit, the landers were supposed to separate from the orbiters, brake their descent through the Martian atmosphere partly by rocket thrusters and partly by parachutes, and touch down gently on the surface. *Mars 2* was launched flawlessly on May 19, and *Mars 3* was away on May 28.

Thus, three spacecraft, one American and two Russian, traversed their transfer orbits across interplanetary space, taking about five months to reach Mars. The American spacecraft arrived in mid-November, some two weeks ahead of the first of the Russian vehicles.

MEANWHILE, however, fateful events had been taking place on Mars itself. Although they took most astronomers by surprise, these developments were not entirely unanticipated. In February 1971, Charles F. Capen at Lowell Observatory had made a prediction in an article about the dust clouds, or "yellow clouds" as they were still known at that time, because they show up best when Mars is observed with a yellow filter:

> Though yellow clouds have been recorded in all Martian seasons, the largest outbreaks seem to occur during Martian perihelic oppositions, when the insolation is greatest on the planet and the thermal equator is far south of the geometric equator. . . . If a bright yellow cloud again develops in the Hellespontus region [as was the case in 1956], . . . it will likely do so after opposition. A vast atmospheric disturbance could interfere with . . . the first Mariner orbiter spacecraft mission, which is planned to begin reconnaissance of the planet in November.[2]

The actual events followed Capen's prediction to an uncanny degree. On September 21, six weeks after opposition, the great dust storm of 1971 began unobtrusively with a small yellow cloud over Hellespontus. One of the first observers to see it was Alan W. Heath, an English amateur using a 12-inch (30-cm) reflector. The storm expanded rapidly. In the 6-inch (15-cm) reflector that was my usual instrument at the time, the dark markings still showed their normal appearance in late September, but by early October they had become ill-defined as a continuous belt of dust clouds spread through the mid-southern latitudes. Within three weeks the entire planet was covered.

On November 10, *Mariner 9* had drawn to within 800,000 kilometers of Mars and switched on its television cameras. The planet remained hopelessly obscured, and the first pictures showed no detail whatever except for the bright south polar cap at the bottom of the disk and four mystifying smudgy spots near the equator.[3] The situation could hardly have seemed worse: the spacecraft had traveled all the way to Mars, only to be clouded out. (It was even suggested in some quarters that the storm was not coincidental, that the Martians were hiding from the spacecraft cameras.)

Under the circumstances, it was indeed fortunate that *Mari-*

ner 9's mission plan had been kept flexible in order to allow for last-minute changes. On November 14, after slowing itself down by firing its thrusters, the spacecraft entered a closed orbit around Mars inclined some 65° to the equator and shut off its television cameras in order to conserve energy until the storm cleared.

On November 27, *Mariner 9* was joined by the first of the Russian spacecraft, *Mars 2*. Shortly before entering orbit *Mars 2* released a descent module, but the lander failed to function properly and crashed onto the surface. Its sole claim to fame is that in doing so it became the first man-made object to reach the surface of Mars (at a site just north of Hellas: latitude 44.2° s, longitude 313.2° w), where it deposited a pennant bearing the insignia of the Soviet Union. *Mars 3* went into orbit on December 2. Again a descent module was released, and this time the capsule actually reached the surface safely, but it remained intact just long enough to turn on its television camera—within a few seconds, contact was lost. At the time it was suggested that the lander might have been blown over by the galelike winds scouring the surface.

The two Russian orbiters fared little better than the landers; they had been preprogrammed to carry out their imaging sequences automatically, and consequently could not wait out the dust storm as *Mariner 9* was doing. Heedless of the dust, they sent back a series of blank and completely uninformative images. But other instruments on board did succeed in sending back some useful information, including measurements of the temperature at various points on the surface—the coldest point proved to be the north polar cap, where the temperature was −110°C (−166°F); elsewhere the values ranged from −93° to 13°C (−135°–55°F), depending on the latitude and time of day.

OF THE flotilla of five spacecraft that had set out for Mars in 1971, all hopes had thus come to center on *Mariner 9*. A month after the spacecraft had entered Martian orbit, the pall of dust had cleared sufficiently for the systematic mapping of the surface to begin. The large dark spots that had at first defied explanation were the first features to emerge from the dust, and proved to be huge shield volcanoes. The largest, with a complex summit caldera, was in the position of Schiaparelli's Nix Olympica (the Snows of Olympus); it has since been renamed Olympus Mons. With its summit towering 25 kilometers above the surrounding plains, Olympus Mons is the tallest mountain in the solar system; since it measures 600 kilometers at the base, however, its slope is very gradual. (By comparison, the largest shield volcano on Earth, Mauna Loa, measures only 120 kilometers across at its base, and the summit rises 9 kilometers above the ocean floor.) The other great volcanoes in this part of Mars are known as the Tharsis Montes—Ascraeus Mons, Pavonis Mons, and Arsia Mons (corresponding in position to the telescopic patches Ascraeus Lacus, Pavonis Lacus, and Nodus Gordii). They are spaced about 700 kilometers apart and aligned southwest-northeast along the crest of the great rise known as the Tharsis bulge; their summits reach 17 kilometers above the surrounding plains.

Mariner 9's first mapping cycle took place predominantly over the southern hemisphere—the region between 25° and 65° s. The most prominent features seen in this cycle included Hellas and Argyre, which are great impact basins. That this is ancient, heavily cratered terrain had been known since the flyby Mariners; but some completely unexpected features emerged as well. *Mariner 9* found networks of channels and tributaries that looked for all the world like runoff channels and dry riverbeds, and which strongly suggested that conditions on Mars must once have been very different from what they are today; once, running water had existed on that surface. The next mapping cycle included regions as far north as latitude 25° N, and revealed the enormous canyon system known as Valles Marineris—the Grand Canyon of Mars as it has been called, though it much bedwarfs its terrestrial counterpart. Valles Marineris extends along the equator for 4,000 kilometers, one-fourth of the way around the planet! Its origin is close to the summit of the Tharsis bulge, at Syria Planum, where it consists of a series of short, deep gashes intersecting at all angles known as Noctis Labyrinthus. In its middle section the canyons become more continuous and run in parallel as three main branches (Ophir, Candor, and Melas Chasmata), which are separated by intervening ridges. The combined width across all three canyons reaches 700 kilometers, and the depth, in places, is as much as 7 kilometers. These canyons connect with Coprates Chasma, which runs eastward and joins Eos Chasma, which finally merges with the large area of blocky "chaotic terrain" near the classical dark area Margaritifer Sinus, which is associated with the large Ares, Tiu, and Simud outflow channels, of which more later.

Other important results included the discovery of the etched, pitted, and laminated terrain around the south polar cap, which consists of sediments overlying an older crater topography and eroded chiefly by winds—and at some past stage possibly by flowing ice. This again seemed to attest to past cycles of climatic change. The spacecraft also returned the first close-up images ever taken of the two Martian moons. By the time *Mariner 9* finally ran out of fuel, on October 27, 1972, it had obtained 7,239 images. The red planet was revealed as never before, and it was neither another Earth nor another Moon; it was Mars—"itself alone."

I DO NOT propose to go into great detail about Martian geology, a vast subject that has been ably treated elsewhere.[4] However, I must at least summarize some of the main points that emerged from a close study of the *Mariner 9* images.

First, with few exceptions, the boundaries of the classical albedo features do not correlate at all well with Martian topography (among the exceptions are the impact basins Hellas and Argyre, which had been well known from Earth as large, circular bright regions). For example, instead of being a dried-up sea basin as was long believed, Syrtis Major turns out to be an elevated plateau; on the other hand, some of the other dark areas, such as Mare Acidalium, are relatively flat plains. The low areas of the planet—the large impact basins—are bright, but so too are the highest, the Tharsis and Elysium rises. In short, the Schiaparellian nomenclature that had long served for the albedo features would simply not do for the topographic features, and it became necessary, in 1973, for the International Astronomical Union to introduce some new terms:

Catena: chain of craters
Chasma: canyon
Dorsum: ridge
Fossa (pl. *fossae*): long, narrow valley(s)
Labyrinthus: intersecting valley complex
Mensa (pl. *mensae*): flat-topped elevation(s)
Mons (pl. *montes*): mountain(s)

Planitia: plain
Patera: shallow crater with scalloped edges
Tholus: small, domical mountain or hill
Vallis (pl. *valles*): valley(s)
Vastitas: widespread lowlands

Broadly speaking, the Martian surface is divided into two main regions. This division has been referred to as the great "crustal dichotomy." South of a circle inclined by roughly 35° to the planet's equator are ancient, heavily cratered highlands; north of this circle are younger, relatively smooth plains and volcanic features. The boundary between the two regions is formed by a gentle, irregular scarp and low, knobby hills. On average, the southern highlands are some 2.1 kilometers higher in elevation than the northern lowlands.

The southern highlands, which include most of the subequatorial parts of the planet as well as a rather wide tongue extending northward beyond Sinus Sabaeus–Sinus Meridiani, are profuse with craters, so much so that the view looks superficially very much like that of the highlands on the Moon—thus the discouraging results of the flyby Mariners. In the case of the Moon, there was a long and heated debate about the origin of these craters; on one side were those who believed them to be impact features, on the other were those who maintained that they had been formed by internal processes of some kind—broadly speaking, by volcanism. During the 1960s and early 1970s, the question was finally settled decisively in favor of the impact theory, and there can be no doubt that the Martian craters were formed in the same way.

The impact process has now been worked out in considerable detail. When an object—say, a small asteroid—plunges into the surface of a planet, it produces two interacting shock waves. The first shock wave engulfs the asteroid, vaporizes it, and melts rock at the immediate point of impact. This part of the process absorbs a relatively small fraction of the energy of impact; the much greater share goes into producing a second shock wave, which travels radially away from the point of impact, excavates the crater, and throws a rim of disintegrated material around it (the ejecta blanket).

Craters of different diameters have different forms. Very small craters are simply bowl-shaped pits that have a fairly constant depth-to-diameter ratio of about 0.20. Larger craters are more complex.

Violent rebound of the floor from the shock of impact gives rise to a central peak or peaks. In addition, many of the larger craters have terraced walls caused by landslip—the slumping in of rim materials toward the center of the crater. This partial filling in with wall material explains why the more complex craters become shallower with increasing diameter.

Even after the flyby Mariners it was obvious that in general the Martian craters are flatter and more subdued than their lunar counterparts, and usually lack the latters' hummocky surrounds. There is also a relative paucity of smaller pits on Mars (less than about 20 km across). These findings are readily explained as being due to the fact that on Mars, unlike the Moon, there has been considerable weathering over time by water, air, and (possibly) glacial erosion. There are other important differences between Martian and lunar craters as well. Summit pits on the central peaks of Martian craters are much more common than on the Moon, and though smaller pits have lunarlike ejecta blankets that have been laid down with ballistic trajectories, those with diameters greater than about 5 kilometers tend to have a different pattern of overlapping sheets of ejecta with lobate margins—the latter being telltale signs of formation by flow across the surface. It has been suggested that the 5-kilometer diameter transition between ballistic and flow patterns may correspond to the minimum depth that needs to be excavated in order to release subsurface water ice.[5]

Martian impact craters with diameters greater than about 50 to 70 kilometers are referred to as basins. The very largest basins, greater than 300 kilometers or so across, are multiringed features similar to those already known on the Moon, though again, because of erosive forces the condition of the Martian features is much less pristine.

The Hellas basin, which measures 2,300 kilometers from rim to rim, is the most imposing topographical feature of the Martian southern hemisphere. Its floor is the lowest point on Mars, located some 5 kilometers below the Martian datum. (On Mars, of course, there is no true "sea level"; instead, the reference datum is defined as the level at which the partial carbon dioxide pressure is 6.1 millibars, which marks the triple point of water. For partial pressures of water greater than 6.1 millibars, liquid water can exist under some temperature conditions; for partial pressures lower than this, it is always unstable.)[6] In the winter Hellas is often covered with carbon dioxide frost, and at such times can appear brilliant white. Other basins clearly visible in the *Mariner 9* images are Isidis and Argyre—the latter measures 1,900 kilometers across and still shows its rim mountains, the Nereidum and Charitum Montes. This basin, too, is often covered with frost. Another major mountain range, the Phlegra Montes, is located in Elysium; these mountains have been identified as remnants of another basin rim that was later inundated by volcanoes. In recent years a number of other large basins have been identified, most of them well-nigh obliterated by Martian weathering.

When the frequencies of craters of different sizes are plotted for the highlands of the Moon, Mercury, and Mars, the resulting distributions, although not identical, demonstrate the existence of two distinct cratering populations: an older population, made during a period of heavy bombardment in which the great basins and the majority of the craters of the rugged highlands were formed; and a younger population that may be traced in the more recent plains and was formed by post–late heavy bombardment.

When the solar system began to form 4.6 billion years ago, a rotating disk of gas and dust began, grain by grain, to accrete into larger objects known as planetesimals; these in turn accreted into the planets. Some of the planetesimals became rather large in their own right and collided with the planets, with fateful consequences. For instance, a smash-up involving a Mars-sized body and the Earth is believed to have given rise to the Moon. Another collision between a large object and Mars about 4.2 billion years ago may well account for the Martian crustal dichotomy. In the low-lying plains of the northern hemisphere, which are now largely covered with sedimentary debris, a gigantic impact basin has been tentatively identified. The Borealis basin, as it is called, is 7,700 kilometers across and is centered in Vastitas Borealis (50° N, 190° W); it is indeed vast—altogether it covers some 80 percent of the northern hemisphere plains.[7]

Obviously space was much more crowded in the early history of the solar system than it is now, and impacts were more frequent. The residue of accretional material subjected the Moon and planets to a massive late bombardment in which the impacts occurred at such high rates that their surfaces became saturated with craters—in other words, the formation of new craters could take place only by obliterating preexisting ones. In the case of the Moon, this violent period came to an end about 3.8 billion years ago. It may have ended at about the same time on Mars, although there is evidence to suggest that it continued until somewhat later. In any case, the objects derived from accretional debris became extinct. Henceforth, craters were added at a much lower rate and were caused by the occasional impacts of asteroids and comets (for Mars, which lies very near the asteroid belt, it has been estimated that asteroids create seven times more craters than comets do). It is these comets and asteroids that are responsible for the post–late heavy bombardment cratering.

Various areas on the Martian surface show widely different cratering densities; the most heavily cratered areas are the oldest. Thus the relative ages of surface units (stratigraphic relationships) can be worked out, and by making certain assumptions about when the late heavy bombardment phase ended and the rates of cratering since, one can even go so far as to estimate their absolute ages (though really reliable values must await the return of actual surface materials from Mars). The oldest units make up the so-called Noachian system, which is represented by the ancient cratered terrain centered on Noachis Terra (approximate ages 4.60 to 3.80 billion years). Overlying the rocks of Noachian age are those of the Hesperian system, whose units are characterized by ridged plains material, of which examples are found in Hesperia Planum and Vastitas Borealis (approximate ages 3.80–3.55 billion years). Finally, the Amazonian system consists of largely smooth plains material such as that which covers Acidalia, Amazonis, and Elysium Planitia (ages less than 3.55 billion years).

AREOLOGISTS ONCE believed that Mars was cool early in its history, and that it formed a hot core at a relatively advanced stage, after heating due to decay of radioactive materials warmed it sufficiently to initiate melting of rock. This would have implied a late onset of volcanism. We now know otherwise, and from an unimpeachable source: through direct analysis of material from Mars itself.

This material exists in the form of several unusual meteorites — one fell at Chassigny, France, in 1815; others fell at Shergotty, India, in 1865 and at Nakhla, Egypt, in 1911; and several others have been identified as well. Their Martian origins have, however, been suspected only since 1981. The meteorites are classified into three groups: shergottites, nakhlites, and chassignites (collectively known as SNCs). All are of the common stony type, but they are very young compared with the 4.5-billion-year age of most meteorites. Gases captured in shock-generated glassy nodules within them were analyzed and proved to have the exact composition of the same gases in the Martian atmosphere; they also contain small amounts of water and water-altered minerals. There can be little doubt that they are Martian. They were blasted into space in one or (more probably) several impacts, with such force that they reached escape velocity and eventually reached Earth.

Detailed analyses of the SNC meteorites made it clear that not long after Mars accreted into a world, its interior was already hot and had differentiated into a nickel-iron core, mantle, and crust. Much of the heat at this stage must have come from the energy of the impacts themselves. Thus, Martian volcanism began early. During late Noachian and early Hesperian times, melted rock (magma) began to reach the surface. Extensive ridged plains were laid down in areas of the southern hemisphere; the so-called highland paterae also emerged, of which four are located near the Hellas basin and probably were formed in relation to deep-seated fractures produced during the impact that formed it. The best-known example is Patera Tyrrhena (at 23° S, 255° W), which seems to have been a volcano of the explosive type, as its irregular summit caldera (40 km long by 12 km wide) is surrounded by large quantities of ash. The plateau of Syrtis Major Planitia (10° N, longitude 290° W) is another early volcanic region; its activity seems to have begun during postimpact adjustments of the crust around the Isidis basin. The dark materials that cover it originated from a low-relief volcanic shield.

This early volcanic period was characterized by the rapid escape of heat from the interior to the surface. Inevitably, the planet began to cool; as it did, convection of the mantle decreased, the overlying crust steadily thickened, and surface volcanism came to be concentrated in ever more limited regions. For some reason not yet entirely understood, the main volcanic activity came to center on two areas: Elysium and Tharsis. These areas marked the locations of "hot spots," or mantle plumes—places where a column of heated material rose from the mantle. Why there should have been only two such plumes on Mars is not known, but the consequences are obvious enough. Lava flooding occurred in these regions on an enormous scale, producing a domical buildup of material which stretched the overlying crust and produced belts of intense fracturing. Along the equator, Valles Marineris began to open when a series of deep troughs formed, oriented radially to the Tharsis rise; later, these troughs were eroded into spurs and gullies.

In Elysium, the first volcano to appear was Hecates Tholus, a shield structure some 180 kilometers across and 6 kilometers high. The center of the eruptions then shifted some 850 kilometers to the south to form the dome-shaped Albor Tholus. Still later eruptions gave rise to Elysium Mons, a shield volcano some 500 kilometers across at its base and standing some 9 kilometers above the surrounding plain; its summit is marked by a 14-kilometer-wide caldera.

By far the greatest activity took place in Tharsis, to the west. The Tharsis bulge stands out like an enormous hump relative to the ellipsoidal shape that the planet would have were it in equilibrium with itself; this hump extends 4,000 kilometers north-south from the plains bordering Mareotis Fossae to Solis Planum, and 3,000 kilometers east-west from Lunae Planum to Amazonis and Arcadia Planitia. Its average level stands some 8–10 kilometers above the Martian datum. In late Hesperian time, the volcanic activity in the region centered on Alba Patera, which lies on the bulge's north flank in an extensively fractured landscape; the caldera itself shows little vertical relief, but measures 1,500 kilometers across. By early Amazonian time, a fault line running on the northwest flank of the Tharsis bulge had become active, giving rise to the three great shield volcanoes of the Tharsis Montes. Several smaller shields and volcanic domes lie close to the same line—Biblis, Ulysses, and Uranius Paterae, and Uranius, Ceraunius, and Tharsis Tholi. Finally, on the southeastern flank of the Tharsis bulge, some 1,200 kilometers northeast of the Tharsis Montes, arose Olympus Mons, whose slopes make up what is probably the youngest surface on Mars. The eruptions here continued long after cooling of the planet's interior had extinguished active volcanism elsewhere—the most recent may have occurred as little as 300 million years ago.

OF THE many discoveries of *Mariner 9,* by all odds the most exciting was the recognition of valley networks and outflow channels, which can hardly have formed otherwise than by the action of running water at or near the surface. Their existence has provided the strongest evidence that Mars may have undergone major climatic changes over time.

The valley networks are in general the older features. They have tributaries, so that they look very much like dry riverbeds, and they lie almost entirely (about 98 percent) in the heavily cratered highlands of the southern hemisphere—and so must be as ancient as they. They are typically 1 to 2 kilometers wide, and not very long; even including their tributary systems, networks seldom go on for more than a few hundred kilometers. The longest—Ma'adim Vallis (centered at 20° s, 182° w), Al Quahira (at 18° s, 196° w), and Nirgal Valis (north of Argyre at 28° s, 40° w)—range in length from 400 to 800 kilometers. At first it was hoped that the valley networks might be proof that precipitation took place on early Mars—rain, in other words—though more recent studies have shown that they give every indication of having been formed by groundwater sapping due to melting of an ice-rich permafrost.[8]

Some valleys formed after the end of the period of heavy bombardment; for example, the system located on the northern flanks of Alba Patera. It is certainly young, and may well be of early Amazonian age. In every way it looks similar to the fluvial valleys produced by runoff on the flanks of the Hawaiian volcanoes, and like the latter is believed to have been formed by surface runoff (though again produced by a sapping process rather than by rainwater).[9]

Compared with the comparatively tiny valley networks, the outflow channels are features on a grand scale; typically they measure hundreds of kilometers long and tens of kilometers wide. They generally emerge in areas of the surface that have undergone collapse, such as chaotic terrain or canyons. Several outstanding examples—the Ares, Tiu, and Simud Valles—originate in the chaotic terrain in Aurorae Sinus at the eastern end of Valles Marineris; Kasei Vallis extends from Echus Chasma, a canyon just west of Lunae Planum; and the Maja, Vedra, and Bahram Valles all arise from Juventae Chasma, on the opposite flank of Lunae Planum. All of these channels then converge on and disappear into the southern floor of

Chryse Planitia on the eastern margin of the Tharsis bulge. There are also many outflow channels in Elysium, northwest of the volcanic province, whence they debouch into the low-lying northern plains; still others are found in Memnonia, Amazonis Planitia, and on the rim of Hellas.

These enormous channels lack tributaries; some, such as Mangala Vallis in the upland region of Memnonia on the border of Amazonis Planitia, are characterized by sculpted landforms such as teardrop or lemniscate islands. The closest terrestrial analogy to these features is the Channeled Scabland of eastern Washington, in the United States, which was formed at the end of the last Ice Age when much of western Montana was covered by a glacial meltwater lake (Lake Missoula). The lake was held in check by an ice dam lying across northern Idaho; when the dam suddenly broke, it released the captured water in a flash flood that drained the whole Columbia Plateau as far as the Pacific Ocean over a period of several days.[10] There is every reason to believe that the Martian channels were also formed by catastrophic flooding, but on an even more monumental scale.

The outflow channels are generally younger than the valley networks, and all were formed after the period of heavy bombardment; the circum-Chryse channels have been dated to Hesperian times, and Mangala Vallis is of Amazonian age. Many of the channels seem to record multiple flooding events over a long period, and apparently they could even form under present conditions; although liquid water is unstable on the surface of Mars because of the low atmospheric pressure (any that formed there would rapidly boil away), this poses no obstacle to massive floods of brief duration.

What was the source of all the water? It is probable that early in its history Mars had a much more substantial atmosphere than it does now. Over time, the oxygen present bonded with rock, turning it red, and the water seeped down through the meteorite-fractured regolith. In the upper parts of the regolith the water froze, but farther down the temperature was still warm enough for the water to remain liquid, and seas formed deep within Mars, pooling beneath an ice-rich permafrost layer perhaps several kilometers thick. Volcanism such as that in Tharsis caused extensive melting of the permafrost layer, releasing the water through permeable volcanic rocks, out along the great fracture systems such as Memnonia

Fossae and Valles Marineris, and finally onto the surface in vast catastrophic floods.

THE IMPLICATIONS of the valley networks and outflow channels are still being debated, and I shall have more to say about them in the next chapter, but there can be little doubt that their discovery was the single most important revelation of *Mariner 9*. To astronomers, geologists, and laypersons alike they suggested the distinct possibility that Mars, in having once allowed running water on its surface, may not always have been so forbiddingly severe as it is now. This in turn gave a tremendous impetus to the next American mission, Viking, whose lofty goal was nothing less than to commence the in situ search for evidence of life on Mars.

The odds of finding living organisms on Mars were obviously very slim, but it seemed that they might not be nil. Such a search was not without its chimerical aspects, but it had been all but inevitable ever since Percival Lowell had sat at the eyepiece of his telescope and, scanning the little globe of orange-yellow spotted by transverse stripes of color, had savored the thought: *Here there be life!*

Vikings—and Beyond

After the rather dismal landscape revealed by the flyby Mariners, the *Mariner 9* images, "the most exciting ever obtained in planetary exploration,"[1] gave a much-needed impetus to Mars studies. But as successful as it was, *Mariner 9*'s reconnaissance had taken place entirely from orbit, and the one thing that it had not been able to do was search for life on the surface. We pass quickly over several Russian spacecraft, including two landers, which sent back depressingly meager results in 1974,[2] and turn to the events of 1975–76, a year that will forever be remembered in the annals of Martian exploration.

NINETY-EIGHT YEARS after Schiaparelli made his memorable telescopic study of the planet and gave the surface features their enchanting names, including Chryse, the Land of Gold, the great attempt began. On August 20 and September 8, 1975, the two Viking spacecraft were launched toward Mars from Cape Canaveral, Florida. Though it was far from being their only purpose in going to Mars, the search for life certainly became the dominant theme of the Viking missions (fig. 20).

Each spacecraft consisted of an orbiter and a lander component. The orbiters were equipped with a more sophisticated television camera system than the one that had flown aboard *Mariner 9,* and each carried instruments for measuring the composition, pressure, temperature, and water vapor content of the Martian atmosphere. Their main purposes, however, at least initially, were to scout suit-

FIGURE 20. An evocative spacecraft view of Mars, created from Viking orbiter images by Alfred McEwen of the U.S. Geological Survey in Flagstaff, Arizona. This view is dominated by the vast canyon system of Valles Marineris. The three large shield volcanoes at the extreme right are, from bottom to top, cloud-filled Arsia Mons, Pavonis Mons, and Ascraeus Mons. (Courtesy U.S. Geological Survey, Flagstaff, Arizona)

able landing sites and then to provide a radio link between the landers down on the surface and the Earth. Only after these tasks were completed could the orbiters be spared to carry out other investigations.

Unfueled, each lander weighed about 600 kilograms. In addition to a miniature laboratory needed to carry out the all-important biological experiments, the landers carried television cameras and an array of sensitive meteorological and seismological devices.

Prospective landing sites had been chosen well in advance. Lander 1 was to touch down at the point at which one of the largest Martian outflow channels (Maja Valles) debouched into Chryse

Planitia, since this location was thought to offer the best hope of finding water and near-surface ice—and hence organic molecules, if any happened to be present. Lander 2 was to land at Cydonia, near the point of maximum extent of the north polar cap, again a favorable site for water.

The journeys across interplanetary space were uneventful. *Viking 1,* actually the second Viking launched, was inserted into Martian orbit on June 19, 1976. Its orbit was an ellipse with a periapsis of 1,513 kilometers and an apoapsis of 33,000 kilometers; each revolution was completed in 24.66 hours—the period was synchronized with the planet's rotation so that the orbiter could assume a stationary position over the intended landing site. Originally the mission planners had hoped that the first landing could be attempted on July 4, 1976, the two hundredth anniversary of the signing of the American Declaration of Independence, but the orbiter photographs showed that the Chryse site was more rugged than expected. There was no choice but to look elsewhere, and by the time an acceptably smooth location had been found—it lay farther to the west, though still within Chryse Planitia—the landing had to be delayed until July 20 (coincidentally, the seventh anniversary of the first Apollo lunar landing).

Early that morning—1:50 A.M. at mission control at the Jet Propulsion Laboratory in Pasadena, California—the *Viking 1* lander, enclosed within a cone-shaped aluminum aeroshell, separated smoothly from the orbiter and began its descent toward the surface. For part of its descent the spacecraft would be traveling at 16,000 kilometers per hour, and since Mars, unlike the Moon, has an atmosphere, air friction could not be ignored—thus the capsule was equipped with a heat shield. In the later stages of descent, when the lander had reached an altitude of 6 kilometers above Mars, a parachute was deployed. At this point the aeroshell was jettisoned and the legs of the lander were extended. Less than a minute later, when the craft had reached an altitude of 1,200 meters above the ground, the parachute was released and retro-rockets were fired to slow the craft for the rest of its descent. Just before it came to rest on the surface, the lander was traveling at a speed of only 10 kilometers per hour.

Scientists on Earth did not know its fate for another nineteen minutes, the time required for the radio signals to travel the 342 million kilometers between Mars and the Earth. Fortunately, the lander narrowly missed a boulder large enough to cause a crash and settled safely on the surface—the precise location was at 22.5° N, 48° W, in Chryse Planitia. At the resolution of the orbiter images, the surface had appeared featureless apart from impact craters and wrinkle ridges—there was little, indeed, to distinguish it from one of the wide gray plains of the Moon.

Within twenty-five seconds of the lander's arrival, one of the television cameras began to scan the surface. The first picture took five minutes to scan—and then, of course, another nineteen minutes to transmit from the orbiter relay link to Earth. The field of view, which lay only 1.5 to 2 meters away from the camera, was strewn with rocks up to 10 centimeters across. The surface was obviously fairly firm; part of one of the footpads was visible, and it had penetrated less than 4 centimeters into the Martian soil.

Immediately after sending back this image, the same camera began scanning a 300° panorama to show the landscape around the spacecraft. Again the image showed rocks of various sizes, most of them angular in shape and coarsely pitted—the whole area was evidently a lava field, and the rocks appeared to be basaltic. It was an exhilarating moment; viewers back on Earth felt as if they were actually standing on the Martian surface. Another panorama, taken by the second television camera, included the 60° sector that had been hidden from the first camera by the body of the spacecraft; this showed an almost Saharan expanse to the northeast with large dune drifts of fine-grained materials. Obviously, aeolian processes were important.

Telescopic observers have long been fascinated by the compelling, if partly illusory, colors of Mars, and the Viking imaging team also had a hard time getting the Martian colors right. They made colored pictures by mixing images obtained through a tricolor wheel (a wheel containing blue, green, and red filters), but the calibration was uncertain, and the earliest published images showed the Martian surface as a very piquant orange-red.[3] Meanwhile, the Martian sky—surprisingly bright and rather unromantically described as "similar to a smoggy day in Los Angeles"—was given a bluish cast that was eerily Earthlike.[4] Later, someone suggested that Mars's atmosphere was probably too thin for molecular (Rayleigh) scattering to produce a blue sky color, and that instead the sky brightness must be due to a suspension of fine reddish dust, which would make the sky salmon pink—and so it became in later images (although I must admit that I never found the effect quite believable).[5] In this case, the Viking imaging team seems to have bent too far in the other direction to avoid creating the illusion of an Earthlike Mars. Because the dust load of the Martian atmosphere is extremely variable, the color of the sky ranges, at different times, from salmon pink to yellow to light blue to dark blue to purple. Since the images returned by the Viking lander just after it landed show shadows cast by the rocks as razor sharp, the atmosphere was then fairly clear. Ironically, after all the trouble taken with it, the sky seems to have been blue after all!

AFTER THE excitement of viewing the first images of the Martian surface subsided, attention turned to the all-important biological experiments. The question of how to detect organisms on Mars, should they exist, had been carefully considered by the biologists on the Viking team (alas, Lowell's notion of large-brained Martians had long ago been discarded; no one expected to find anything more sophisticated than lowly microbes). The biologists designed three experiments, described below, to test for the presence of life.

1. *Pyrolytic release experiment.* A tiny sample (0.1 g) of Martian soil is first scooped up with the lever arm of the lander, then placed inside the test chamber. Carbon dioxide and carbon monoxide labeled with the radioactive isotope carbon-14 are then admitted, the mixture is incubated under a sunlamp for several days, and everything is heated in order to break down (pyrolyze) any organic compounds present. Finally, hydrogen gas is admitted into the chamber to sweep the pyrolysis products into a gas chromatograph and mass spectrograph capable of detecting carbon-14. Since any organisms present should carry out metabolic processes during which they will assimilate carbon-14 from the gas in the chamber, detection of carbon-14 would be a positive result, though in and of itself not entirely conclusive, since a first peak of radioactivity might equally well be due to chemical processes not involving living organisms. In order to rule out this latter possibility, other samples (serving as controls) are sterilized by heating before the carbon source is admitted.

2. *Labeled release experiment.* Again, a sample of Martian material

is placed into the chamber, and a moist nutrient material containing carbon-14 is added. Any Martian organisms present will metabolize the nutrient material and release carbon-14-labeled gas, which is then registered by the detector.

3. *Gas exchange experiment* (popularly known as the "chicken soup" experiment). At the beginning of the experiment, the atmosphere within the chamber consists of carbon dioxide and the inert gases helium and krypton; a nutrient material and water vapor are added to the soil sample. On suddenly finding themselves in a water- and nutrient-rich environment, the Martian organisms will respond with a vigorous spurt of metabolism, resulting in the sudden buildup of gases in the chamber.

Apart from these three experiments, the mass spectrograph was designed to make sensitive measurements in a direct attempt to detect organic materials on Mars.

Though straightforward enough in theory, the experiments produced confusing results. The pyrolytic release experiment showed two peaks, and at first was felt to be weakly positive; however, later attempts to duplicate the effect were unsuccessful. The labeled release experiment showed an immediate—and startling—rise in the level of carbon-14 radioactivity immediately after the nutrient was introduced into the chamber. This data strip seemed to attest to a positive reaction, so much so that the experiment team immediately rushed out and ordered a bottle of champagne. The "chicken soup" experiment also produced dramatic and unexpected results; when the samples were humidified, there was a sudden burst of oxygen—something that had never occurred in earlier tests with terrestrial samples. However, there was a very weak response when the nutrient material was added.

On the whole, the consensus was that, instead of being produced by organisms of some kind, the chemical reactions observed were entirely due to a highly oxidizing substance in the rust red Martian soil, perhaps iron peroxide or superoxide. Some of the organic compounds in the nutrient would have been sensitive to oxidizing materials. The same exotic chemistry can explain the evolution of the large amounts of oxygen when water vapor was added to the gas-exchange experiment. The fact that the mass spectrograph failed to identify any organic compounds (except for some known contaminants such as methylchloride and freon-E) would appear to argue rather strongly against biological explanations for the results.[6] On the other hand, the data from the labeled release experiment remain in dispute, and a small minority of scientists still maintain that the Viking results are positive evidence for the existence of microorganisms on Mars. The last word remains to be said.[7]

SO MUCH for the results at the *Viking 1* site. Meanwhile, of course, *Viking 2* was on its way; it arrived in Martian orbit on August 7, but because the *Viking 1* experiments were then in full swing, the landing was delayed until September 3. Again there were last-minute changes. The original site in Cydonia was too rough, and the lander was directed to an alternate site at Utopia Planitia, on the vast northern plains. Despite a momentary power shortage in the orbiter, which caused a temporary loss of the main communications link, the *Viking 2* lander touched down safely at 48° N, 225.7° W, some 7,400 kilometers northeast of where the *Viking 1* lander was already resting on the surface.

As the first panoramic view began to come back from Utopia, the horizon appeared strongly sloped to the right; one of the lander's footpads had come to rest on a boulder, canting the spacecraft at an 8° angle. Though the orbiter photographs had suggested that this would be a region of dunes, Utopia was no less a forest of rocks than the *Viking 1* site had been. There were differences, of course. Unlike the rather varied rock forms of Chryse, Utopia's rocks proved to be larger on average and more evenly distributed across the surface, with no bedrock outcrops or large drifts, and this created a singularly monotonous appearance. The most striking feature about the rocks, apart from their uniform size and distribution, was their extensive pitting. They looked rather like terrestrial vesicular volcanic rocks, in which pits form when cavities are created around small gas bubbles in volcanic lava.

Whereas the Chryse site consisted of gently rolling plains, Utopia proved to be remarkably flat, presumably because of its proximity to the 90-kilometer impact crater, Mie, whose rim lay 170 kilometers east of the landing site. The orbiter photographs showed that a broad lobe of ejecta material from Mie ran just southeast of the site. Aeolian processes were again evident, but the drifts were smaller than in Chryse—they occurred only in patches between clusters of rocks and as small windtails. Finally, the same suite of biological experiments was carried out at the *Viking 2* site, and they too failed to find conclusive evidence for or—for a minority of scientists—against the existence of organic life.

THE LANDERS had originally been designed to function for only ninety days. In fact, all four spacecraft—the two landers and the two orbiters—continued to perform much longer than expected. Orbiter 2 transmitted data until July 25, 1978; Lander 2 (VL2) until April 11, 1980; and Orbiter 1 transmitted until August 7, 1980. The longest record was provided by Lander 1—later renamed the Mutch Memorial Station (MMS) in honor of geologist and landing team leader Timothy Mutch, who fell to his death in 1980 while climbing in the high Himalayas; it remained in contact until November 13, 1982, a total of 6.4 Earth, or 3.4 Martian, years.

Apart from the data obtained in the biology experiments, the landers gathered a tremendous amount of useful information about Martian surface conditions. Among other things, they deployed meteorological booms with temperature, pressure, and wind sensors, which effectively served as in situ meteorological stations for three Martian years.

When the landers first arrived, it was early summer in the northern hemisphere, and the atmosphere was nearly dust-free. The diurnal cycle of temperature variation repeated very nearly from one Sol to the next; at MMS, the diurnal temperature range was 50°C, with a minimum at dawn of –83°C (–118°F) and a maximum in the early afternoon of –33°C (–28°F). (Remember, this was the Martian tropics!) The temperatures at the more northerly VL2 site were five to ten degrees colder. Incidentally, because the thin Martian air has virtually no capacity to hold heat, the air temperatures were some twenty degrees colder than the surface temperatures.

OBVIOUSLY, then, Mars is a very cold place—Martian temperatures are a far cry from the south of England temperatures that Percival Lowell predicted. The landers' temperature measures were supplemented by those of the orbiters. The warmest temperatures, found in some of the southern hemisphere "oases," reached as high as 22°C (60°F) in early afternoon in midsummer, but even so, the temperature still plummeted to –53°C (–63°F) at night.

The atmospheric pressure at MMS in the summer was 6.7 millibars; at VL2, because of its lower elevation, the pressure was slightly higher, 7.4 millibars. Winds at both sites were very light, with speeds of 2 meters per second at night and up to 7 meters per second during the day.

Later, the two stations recorded significant seasonal changes. By mid-autumn the polar hood had developed over the north polar cap and storms began passing regularly north of the landing sites, causing fluctuations in pressures and winds, especially at the more northerly VL2 station. In early 1977, a major dust storm occurred; its onset was heralded in orbiter images taken in February that revealed an extensive cloud swirling about the high ground of Claritas Fossae, west of Solis Planum. Within a few days this storm had grown to global proportions. The landers recorded that during the dust storm the Martian atmosphere became much more opaque; the diurnal temperature range narrowed sharply from fifty degrees to only about ten degrees, and the wind speeds picked up considerably—indeed, within only an hour of the storm's arrival at MMS they had increased to 17 meters per second, with gusts up to 26 meters per second. Nevertheless, no actual transport of material was observed at either site, only a gradual brightening and loss of contrast of the surface material.

The dust began to clear slowly over the next several weeks, and as it did so, a thin coating of water ice formed on the rocks at the VL2 site.[8] In May 1977 a second global dust storm developed; the regular dust-free seasonal cycle of pressure and temperature variations was not reestablished until late 1977.

The atmospheric pressure at the two sites also showed seasonal variations. The values climbed from 6.7 millibars at MMS and 7.4 millibars at VL2 in summer to 8.8 millibars and 10 millibars in winter. This behavior had been expected on the basis of the behavior of the seasonal polar caps. Recall that the Martian atmosphere is 95 percent carbon dioxide, and that seasonal caps form at the poles when carbon dioxide freezes out as a frost onto the surface. Because of the longer southern hemisphere winters, the southern seasonal cap becomes much more extensive than its northern counterpart and incorporates more carbon dioxide from the atmosphere; conversely, during the short but hot southern hemisphere summer, much of this carbon dioxide sublimes away again to produce a net re-release of large amounts of carbon dioxide into the atmosphere.

Some 30 percent of the Martian atmosphere is cycled through the seasonal polar caps in this way, and this is bound to have a dramatic effect on the seasonal wind patterns.

It is worth reviewing the different behavior of the two caps in some detail. The south seasonal cap, at its maximum extent, is roughly circular around the pole and reaches to about latitude 60° s throughout most of its circumference (and even somewhat farther, to about 50° s, within the Argyre basin; among other notable outliers is the great Hellas basin, which during the winter is filled with carbon dioxide frost). As the cap retreats, it develops rifts and becomes irregular in outline. Near midsummer, a large peninsular section, located at about 70° s and 330° w, breaks off, forming the detached island of Novissima Thyle (as Schiaparelli called it), or the Mountains of Mitchel. The spacecraft photographs show no mountains, however, only a southward-facing slope on which the frost continues to linger; it consists partly of the scarp of a large impact basin centered some 6° from the pole. The further retreat of the cap reveals an underlying residual cap of carbon dioxide ice, which persists throughout most summers. Its highly characteristic "swirl" pattern is produced by the uneven removal of frost by wind along the slopes of valleys. As long as the residual cap remains, the temperature is buffered to the frost point of carbon dioxide (–125°C [–193°F] at a surface pressure of 6.1 millibars), well below the sublimation point of water.

The northern cap behaves quite differently. When it emerges from the polar hood in early spring, it covers its greatest extent, to approximately 65° N latitude. It too shows rifts and outliers as it begins to retreat, though because the terrain is smoother than in the south there is nothing comparable to the Mountains of Mitchel. Dunes belt the region between 75° and 85° N, forming a collar around the cap. The north seasonal cap is smaller and darker than the southern one—mainly because it is laid down during the part of the Martian year when there is generally more dust in the atmosphere, which precipitates with carbon dioxide frost onto the surface. It also sublimates more efficiently, and vanishes completely by summer solstice, leaving behind an underlying residual cap which, unlike its counterpart, consists not of frozen carbon dioxide but of water ice—one of the surprises of the Viking mission.[9] In all, the amount of water it contains would probably cover Mars to a depth of 10 to 40 meters if it were distributed evenly across the surface. After the carbon dioxide frost disappears, the residual northern cap is no longer buffered to the frost point of carbon dioxide as the southern cap generally is, and in summer may attain a temperature as high as −68°C (−90°F), close to the frost point for water on Mars. These temperatures are consistent with observations of large amounts of water vapor in the northern hemisphere during these times.

THE ATMOSPHERE of Mars is thin, cold, and bone dry. In some ways it resembles the stratosphere of Earth, though with at least one important difference—in the stratosphere, the relative humidity is very low, whereas the atmosphere of Mars is nearly saturated. The saturated water vapor pressures for some Martian temperatures are as follows:

−69°C	0.0031 millibars
−40°C	0.128 millibars
−30°C	0.380 millibars
−20°C	1.046 millibars
−10°C	2.587 millibars

Thus, at −69°C water vapor makes up a mere 0.04 percent of the atmosphere (given a total atmospheric pressure of 715 millibars), and at still lower temperatures the maximum water content is vanishingly small; the result is that the lower atmosphere is easily oversaturated and readily gives rise to clouds.

Water vapor is a minor component of the Martian atmosphere. The main component, of course, is carbon dioxide, which tends to freeze out at higher altitudes into a haze of fine crystals. As already noted, the amount of carbon dioxide varies seasonally by about 30 percent. Based on a total atmospheric pressure of 7.5 millibars, the following values for the composition of the Martian atmosphere have been derived from the Viking lander measurements: carbon dioxide, 95.32 percent; nitrogen, 2.7 percent; argon, 1.6 percent; oxygen, 0.13 percent; carbon monoxide, 0.7 percent; water vapor, 0.03 percent; inert gases, trace.

The atmosphere, thin as it is, can nonetheless transport significant amounts of sand and dust. The Viking orbiters noted many features formed by wind, including vast dune fields, especially near the south polar cap; wind-eroded hills, or yardings; and windblown streaks, both light and dark. Indeed, as noted by V. A. Firsoff, "the dusky areas seen in the distant views of Mars dissolve in the close-ups into swarms of dark 'streaks' and 'splotches.' The streaks are elongated markings 10 or more km long, frequently in the form of 'crater-tails,' which may be comet- or fan-shaped or resemble the flame of a candle, including its central and peripheral shading. Splotches are irregular or rounded, often found within craters, centrally or on the side, and may 'wash over' the walls."[10] The dark streaks ("shredded streaks") that are found in great abundance in some areas, such as Syrtis Major, develop when rocks are scoured by very high winds; the light streaks consist of fine sand blown by the prevailing winds. Clearly, the broad features visible from Earth represent an integrated view of these small details. Changes observed in them, including the seasonal "wave of darkening" (if it exists), may be related, as suggested by Carl Sagan and others, to "the alternate deposition and deflation of windblown dust having detectable contrast with respect to basement material."[11]

THIS DISCUSSION of wind features brings us at last to the famous dust storms of Mars. Dust clouds, stirred up by rising cells of warm air which carry dust from the surface high into the atmosphere, tend to arise preferentially in certain areas; in the southern hemisphere, the active areas include the circumference of the south polar cap, Hellas, Hellespontus-Noachis, and Claritas Fossae–Solis Planum; in the northern hemisphere, some active areas are Chryse-Acidalium, Isidis Planitia–Syrtis Major, and Cerberus. Usually the dust arises at several points on the planet at once; indeed, a cloud's apparent movement may be partly illusory, because small storms tend to form simultaneously and then coalesce with one another. The storms may remain fleeting and localized, or they may spread through a latitudinal corridor around the entire planet (planet-encircling storms). The largest storms are global events like those of 1956 and 1971. A planet-encircling storm occurred in 1973, and two occurred during 1977 while the Viking landers were on the planet; yet another seems to have been developing in 1982 just as the Viking mission was coming to an end. These massive storms (planet-encircling or global) always occur during southern spring and summer, when Mars is near perihelion ($L_s = 251°$), though the actual interval is rather broad—they have been observed to begin anywhere from L_s = 204° to 310°.

Among the factors critical to triggering the development of major storms, the most important is the fact that there is increased heating near perihelion, which in turn produces stronger winds. Again, the atmospheric pressure is 30 percent greater during the southern hemisphere spring and summer than it is in winter, owing to the release of large amounts of carbon dioxide from the seasonal polar cap—which itself enhances the atmosphere's capacity to carry dust. Finally, since the areas where dust clouds generally develop are areas of slopes—or, in the case of the polar cap, of high temperature gradients—topographic factors further enhance the near-surface winds, which is why local clouds tend generally to arise in the same areas. Once aloft, the suspended dust becomes a major absorber both of solar radiation and of heat being reradiated from the surface; it is certainly a much more effective absorber than the thin, cold Martian air.

Dust storms on Mars are an example of a so-called chaotic phenomenon. Positive feedback mechanisms amplify the initial disturbance, but the interplay of these factors is unpredictable—it may lead to large-scale storms, but this is not the inevitable result. (There must, obviously, be negative feedback mechanisms as well, which dampen out the dust storm activity eventually, but at present these are even less satisfactorily understood than the positive feedback mechanisms.)

In general, dust is transported from the dark highland areas of the southern hemisphere, which contain numerous rocks and outcrops where active erosion is taking place, to the bright "deserts" of the northern hemisphere, such as Tharsis, Arabia, and Elysium, which contain significant deposits of fine dust. The north polar cap is also a major depository of dust.

Even a thin coating of dust is enough to account for many of the long-observed changes in Martian albedo patterns. At the end of the major dust storm season (the southern hemisphere spring and summer), the surface features often display decreased contrast due to the continued presence of fine dust in the atmosphere and the deposition of a thin coat of dust on the surface. By the middle of the southern autumn (northern spring), the atmosphere has cleared again, and classical albedo features such as Syrtis Major—always faint immediately after a global dust storm—have fully redeveloped. Most of the dramatic changes in individual features can be explained thus; for example, marked changes in Solis Lacus, the albedo feature associated with the Solis Planum region, were observed after the 1956 and 1973 storms.

For a time it seemed that a major dust storm must develop every time Mars came to perihelion, but we now know that things are not so simple—indeed, there has not been a planet-encircling storm since 1982, although there have been regional obscurations. It is significant that in 1969, the residual carbon dioxide cap around the south pole seems to have disappeared completely, and large amounts of water vapor were detected during the southern hemisphere summer.[12] These unusually warm conditions preceded the development of the great storms of 1971 and 1973. During the 1980s and 1990s, by contrast, Mars has apparently been much colder; the Hubble Space Telescope showed thin cirrus clouds over extensive areas of the planet in 1995, but virtually no dust. It may well be that the great storms of 1956 and 1971 were highly anomalous events, and that the usual Martian conditions are more like those seen in recent years.[13]

MARS'S ASYMMETRIC polar caps and cycle of southern hemisphere spring and summer dust storms reflect the current position of its axis and the eccentricity of its orbit. At present, Mars's axial tilt, or obliquity—25.2° from the perpendicular—is very nearly the same as the Earth's (23.5°). The current agreement, however, is a sheer coincidence. Both Earth and Mars bulge slightly at the equator because of the centrifugal force of their rotation. The gravitational pull of the Sun on these equatorial bulges causes the axial tilts of both Earth and Mars to vary over time. Earth's axial tilt is largely stabilized by the presence of the Moon, and so ranges through only four degrees. Mars, which lacks a large and massive satellite, wobbles in a much more extreme fashion—at the current epoch, its axial tilt ranges between extremes of 15° and 35° over a period of 120,000 years, with the present value lying close to the mean.[14] The spin axis also wobbles, or precesses, just like a top slowing down, with a period of 173,000 years (compared with 25,800 years for the Earth). This is the effect that on Earth gives rise to the well-known precession of the equinoxes.

The planetary orbits themselves rotate slowly in space, resulting in a gradual shift in the position of the perihelion. As a combined effect of the precession of the spin axis and the advance of the perihelion, alternate poles of Mars tilt toward the Sun at perihelion every 25,500 years—that is, on a 51,000-year cycle. The orbits also change shape over time, and again the more extreme changes belong to Mars—its orbital eccentricity (now at 0.093) ranges between 0.00 and 0.13 over a period of 2 million years, while that of the Earth (now 0.017) never exceeds 0.05.[15]

Periodic oscillations in the obliquity of Earth's axis and the eccentricity and precession of its orbit give rise to the so-called Milankovitch cycles, named for the Serbian astronomer M. Milankovitch, who in 1938 proposed that such cycles may partly explain the Ice Ages. We know, for example, that during the past 3 million years, much of the Northern Hemisphere has been covered with ice, with the last glacial maximum occurring 18,000 years ago—indeed, we may not have fully emerged from it. There have been similar episodes throughout much of the Earth's history. Though other factors may also play a role—for instance, the drift of a continental mass over a pole seems to be a necessary precondition of extensive glaciation, and large impacts too may be important because they can raise large quantities of dust and thus reduce the incident solar radiation—it is generally agreed that the effects of the Milankovitch cycles on Earth's climate are far from negligible. On Mars, which lacks the moderating effect of oceans and suffers much more extreme variations in its axial tilt and orbital eccentricity, they may be even more decisive.

On Mars, the most important cycle of climate change is the 51,000-year cycle caused by the combined effect of the precession of its axis and the advance of its perihelion. Although at present the southern hemisphere is tilted toward the Sun at perihelion, in 25,500 years it will be the northern hemisphere instead. The northern hemisphere will then have the short, hot summers, and since the occurrence of the major dust storms is clearly related to Mars's arrival at perihelion, dust storm activity will presumably shift mainly to the northern hemisphere's spring and summer. The large amounts of fine dust currently deposited in the northern hemisphere in regions such as Tharsis, Arabia, and Elysium will be redistributed to the southern hemisphere, and dust accumulation at the south polar cap will exceed that at the north. The asymmetry of the caps will be completely reversed, and the large seasonal carbon dioxide frost cap will form over the north pole instead of the south pole.

Indeed, there is unequivocal evidence for such climatic cycles on Mars in the layered deposits, or laminated terrain, of the polar regions, discovered by *Mariner 9* in the case of the south pole and well documented in the *Viking 2* orbiter images of the north pole. This laminated terrain covers most of the area beyond the 80° circles of latitude—in the southern hemisphere, it is in a region of ancient cratered terrain; in the northern hemisphere, in an area of smooth plains. The layers are thought to consist of alternate strata of dust and ice (or of dust and ice in varying proportions). Presumably these layers record changes in water and dust transport or removal to the polar regions during different periods.[16]

THE MOST interesting climatic conditions are found at the extreme values of the obliquity. Detailed calculations have shown that the obliquity of Mars has ranged between a low of 13° and a high of 47° over the past ten million years (since the obliquity is chaotic, it is inherently unpredictable over significantly longer periods). At the minimum value, 13°, permanent caps of carbon dioxide must form over both poles, and the planet must go into deep-freeze. With most of the atmosphere taken out of circulation, the surface pressure must drop to less than a millibar, in which case there would be too little air to support dust, and dust storm activity must cease altogether. At the maximum value, 47°, both poles would lose their carbon dioxide ice caps each summer. This, obviously, is the most interesting case, since the polar temperatures, no longer buffered to the carbon dioxide frost point, would rise appreciably. More water vapor would be released into the atmosphere, thereby producing still further warming owing to the greenhouse effect.

The greenhouse effect is a result of the fact that certain gases, such as carbon dioxide and water vapor, are transparent to visible light but absorb strongly in the infrared. Thus the light is able to penetrate through the atmosphere but is trapped when it is reradiated as heat from the ground. This produces a net warming of the planet. On the Earth, greenhouse gases are present in very small quantities—carbon dioxide, for example, represents only 0.0003 of the Earth's total atmospheric mass. However, despite the scarcity of these gases, they play the major role in determining the Earth's temperature. At present the average global temperature is about 15°C, around thirty-five degrees warmer that it would be without the effect of greenhouse warming; though this sounds modest, it is very significant, since without it the Earth's oceans would freeze over. Obviously one can get too much of a good thing—on Venus, for instance, which has a massive atmosphere made up 97 percent of carbon dioxide, the greenhouse effect has proceeded with sinister efficiency; the global temperature is 500°C, so the surface is easily hot enough to melt lead. (This may present a cautionary tale to the Earth, where these gases have been building up gradually over the last hundred years or so owing to human activities on the planet. We certainly don't want to end up like Venus!) On Mars, with its very thin atmosphere, the greenhouse effect (even with the help of dust) has increased the global temperature by a mere seven degrees, to about −50°C, which is far too cold for liquid water to form anywhere on the surface.

When the obliquity of Mars reaches its greatest value, the carbon dioxide polar caps must sublimate completely. At such times, Mars's atmosphere would be much more massive than it is now, so that greenhouse warming might increase the global temperature another 30°C—still not enough to allow liquid water to exist on the surface. And yet, water obviously *has* flowed on the surface of Mars, probably repeatedly, so some other mechanism must be involved. Recall that the valley networks are found almost entirely in the old, heavily cratered Noachian terrain, but the outflow channels are younger, being of Hesperian and Amazonian age. The largest of them debouch onto the wide northern plains, where ponded sediments, shorelines, and other features associated with standing water have been tentatively identified. Apparently the northern plains have been periodically inundated by a great ocean, the Oceanus Borealis.[17] Since changes in Mars's atmosphere don't seem to be sufficient to explain these flooding episodes, there can be little doubt that they have been driven by hydrothermal processes—in other words, by heat generated within Mars itself.

WE NOW know a good deal about current conditions on Mars (see appendix 3), and what we know suggests that it is probably not alive. Its environment is harsh in the extreme: cold, all but airless, dry beyond the driest deserts of the Earth. But of all the obstacles to life as we know it, it seems that the lack of water is probably the most important.

And yet Mars has not always been as dry as it is now. The details remain uncertain; what is clear is that there do seem to have been intervals in which the planet has been warmer and wetter than the usual cold, dry post-Noachian conditions—brief oasislike interludes in which life itself might have gotten a foothold. The most recent flooding episodes, indeed, may have occurred within the past several hundred million years. Inevitably, alas, the water seeped back into the regolith again, to be trapped as groundwater and permafrost; the carbon dioxide, too, precipitated out, and once more cold, dry conditions prevailed.

There is at least the chance that life may have gotten a start during the more benign periods of Mars's history. Though it is all but certain that Mars is not now the abode of life, the questions of its past still haunt us and are much further from solution.

The Hurtling Moons of Mars

Obviously the Viking missions were a watershed in the study of Mars. Since then, three more spacecraft have been to the planet. In July 1988, two Russian spacecraft, *Phobos 1* and *Phobos 2,* were launched. Contact was lost with *Phobos 1* on its way out from Earth, but *Phobos 2* successfully entered Martian orbit in January 1989. During the next fifty-nine days it obtained enough photographs to map nearly the entire planet—unfortunately, the full results have not yet been published in the West. There was also the American *Mars Observer,* which, in a stunning setback, went dead in August 1993, just as it was entering the final phase of its approach to the planet—only three days from its destination!

The primary objective of the Russian Phobos mission had been not the planet itself, but Phobos, the larger of the two Martian satellites. Plans called for placing a small lander on the surface of Phobos, but unfortunately, contact was lost in March 1989, just as *Phobos 2* was starting to image the small moon and approach it for the landing phase.

Indeed, tiny as they are, the moons are intriguing worlds in their own right. The events leading up to their discovery by Asaph Hall in 1877 have already been discussed, but, strangely, their existence had been guessed on several earlier occasions, including by Jonathan Swift in 1726.

That year Swift published *Gulliver's Travels,* which describes the imaginary exploits of Lemuel Gulliver. Though his visit among the tiny Lilliputians is perhaps the best known, Gulliver made other explorations. On his "Voyage to Laputa," Gulliver learns that the scientists there

> have . . . discovered two lesser stars, or satellites, which revolve about Mars; whereof the innermost is distant from the center of the primary planet exactly three of its diameters, and the outermost five; the former revolves in the space of ten hours, and the latter in twenty one and a half; so that the squares of their periodical times are very near in the same proportion with the cubes of their distance from the center of Mars; which evidently shows them to be governed by the same law of gravitation that influences the other heavenly bodies.[1]

Swift's prediction is surprising in that he not only had the number of moons right, but he also placed them close to the planet—the distances of the actual Martian moons are 1.4 and 3.5 diameters of Mars, compared with 3 and 5 as given by Swift. One would almost be tempted to think that Swift obtained an actual glimpse of the moons through a telescope, were it not for the fact that there was no telescope at the time anywhere close to being powerful enough to show them. Voltaire, in his 1750 story *Micromégas,* which tells of the visit by an inhabitant of the star Sirius to the solar system, also credited Mars with two moons, but here, at least, there is no mystery; he must have been influenced by Swift's tale.

The idea that Mars might have two satellites harks back still earlier, however, to Kepler's misconstrual of the anagram in which Galileo announced the discovery of what we now know to be the ring of Saturn.[2] Probably Swift had learned of Kepler's earlier surmise. Moreover, since at the time he wrote it was believed that Mercury and Venus were companionless, Earth had one satellite, Jupiter had four, and Saturn had five, Mars's place in this progression seemed to call for two moons. Since they remained hidden, the moons had to be very small, and if they were very close to the planet they would be lost in its glare. However Swift arrived at his prediction, there can be no doubt that it was simply a lucky guess.

AFTER THE proper discovery of the satellites by Asaph Hall in August 1877, it was immediately apparent that they are highly unusual objects. Phobos lies at a distance of 9,400 kilometers from the center of Mars, or only 6,000 kilometers from the Martian surface. (Mars as seen from Phobos would be an astounding sight; its disk would subtend an angle of 43°, and it would fill nearly half the sky from horizon to zenith!) The present period of revolution of Phobos around Mars is only seven hours and thirty-nine minutes. Thus it completes three full revolutions in the time that Mars takes to rotate once on its axis—a state of affairs so surprising that Hall at first thought there must be two or three inner moons! Owing to its rapid motion, Phobos rises in the west and sets in the east, and it remains above the horizon for only four and a half hours at a time.

Because its orbital inclination is only about 1°, Phobos, for all practical purposes, lies in the equatorial plane of the planet. It is eclipsed by the planet's shadow 1,330 times every Martian year, managing to escape only for brief periods around the times of the summer and winter solstices. Observers on the Martian surface above 70° north and south latitudes would never catch sight of it at all, since it would never clear the horizon.

Deimos lies 23,500 kilometers from the center of Mars, and its orbit, too, is nearly equatorial. The period of revolution is about thirty hours, and it remains above the Martian horizon for sixty hours at a time. It never rises above the horizon in the polar regions above 82° north or south latitude.

In 1945, after analyzing measures of the positions of the satellites made since their discovery in 1877, B. P. Sharpless announced that Phobos appeared to be rapidly spiraling inward toward Mars.[3] Such an acceleration could only be produced by some sort of drag, and in 1959 a Russian astronomer, Iosif Shklovskii, concluded that the drag was due to friction with the outer atmosphere of Mars. This was reasonable enough; however, in order to explain the rapid rate of its acceleration, Shklovskii went further and proposed that Phobos must be hollow inside—and that it might even be an artificial space station![4] Subsequently, someone suggested that the reason the satellites were not discovered until 1877, despite careful searches by William Herschel and Heinrich d'Arrest, was that they did not yet exist!

Needless to say, Shklovskii's view was always regarded with considerable skepticism, and later studies have shown that although Phobos is indeed spiraling inward toward Mars, the rate of its acceleration is only about half that derived by Sharpless—about 15° in orbital longitude since 1877. This is a small enough quantity to be accounted for by frictional forces due to tides raised by Phobos in the solid body of Mars. The acceleration will continue for another 40 million years or so, until the moon immolates itself by crashing into the planet.[5]

Owing to similar tidal forces, Deimos, whose period of revolution is slower than the period of Mars's spin, is spiraling very slowly outward from Mars; however, the effect is very slight and actually produces very little change in its orbit.

BOTH MARTIAN satellites are tiny, and this, together with their proximity to the bright planet, explains why they were not discovered earlier. In Earth-based telescopes they are mere glints of light, and only with the advent of the spacecraft era have we begun to find out what they are really like (appendix 4).

The first close-up pictures of Phobos and Deimos were obtained by the *Mariner 9* spacecraft in 1972; since then, they have also been imaged by the Viking orbiter spacecraft and the Russian *Phobos*, which sent back some useful results from Martian orbit in March 1989 before suddenly losing contact. Phobos, which measures 27 by 19 kilometers, is shaped rather like a potato; Deimos too is oddly shaped, though less so than Phobos, and measures 15 by 11 kilometers.

Both moons have suffered heavy bombardment and have numerous impact craters to show for it. Phobos has a particularly large one, named Stickney (after the maiden name of Asaph Hall's wife, who encouraged him to continue his flagging search for the moons). It is 10 kilometers across, and the impact that formed it must have come close to smashing Phobos into pieces. Radiating in all directions from Stickney are a series of ridges and grooves. The grooves are widest (700 m) and deepest (90 m) close to the crater itself, and they converge again near the crater's antipode, which is nearly groove-free. Obviously these features are intimately associated with Stickney itself, and seem to be deep-seated fractures formed during the impact. After Stickney, the largest craters on Phobos are Hall, Roche, Todd, Sharpless, and d'Arrest.

Deimos's surface appears different because most of the craters are partially filled with debris; in many cases they can be identified only because of their bright rims. The two largest, Swift and Voltaire, measure about 3 kilometers across.

The surfaces of both satellites are quite dark, so they are not very effective for lighting up the lonely Martian nights. From Mars, Phobos would appear only about as bright as Venus does from Earth, and Deimos would resemble the bright stars Vega or Arcturus. The Martian moons are thought to be captured asteroids (or asteroid fragments), and in many ways they resemble the asteroids that have thus far been imaged at close range; 951 Gaspra and 243 Ida even have grooves like those around Stickney. There can be little doubt that they are related kinds of objects.

But if Phobos and Deimos are captured asteroids, the details of their capture remain rather murky. Most asteroids stay within the main asteroid belt, but at 2.5 astronomical units (a.u.) there is a clear zone; asteroids there are in a resonance position with Jupiter—that is, they complete exactly three revolutions for every revolution that Jupiter completes. They are, then, regularly disturbed, and as a result their orbits are chaotic. Their orbital eccentricities can become so great that they can even cross the orbits of the other planets—many cross the orbit of Mars, and a few, known as the Apollo group, veer inside that of the Earth.

Rarely, one of these asteroids might be captured by Mars, but if so, it would first have to lose energy, perhaps through aerodynamic drag. Soon after its formation, Mars may have been surrounded by a nebula; an asteroid passing through this nebula would have been slowed enough by friction for its orbit to decay, first into a closed elliptical path around Mars, and later into a more circular orbit. It would continue to spiral quickly in toward Mars until it reached the point where its period became synchronous with the rotation of the planet, after which there would have been little relative velocity between the captured object and the nebula. At this point it would have been stabilized. This would have occurred early in the history of the solar system, when space was still cluttered with rubble. An impact with a stray object later may have broken the synchronous moon apart—the fragment which then became Phobos landed inside the synchronous position, and owing to tidal forces has continued to spiral inward ever since, while that which became Deimos landed outside, close to its present position.[6]

This is plausible enough, but is it true? At the moment we simply do not know; it remains equally possible that the satellites are planetesimals left behind within Mars's gravitational sphere of influence after the planet itself was formed—examples of the kind of objects whose impacts on Mars created the Hellas and Argyre basins during the violent bombardment of the Noachian Age.

We still have a great deal to learn about the Martian moons, but it is sobering indeed to realize that we now have detailed maps of the surfaces of these objects, which for almost a century after their discovery appeared in even the largest telescopes as mere specks of light.

Observing Mars

Mars is only half the diameter of the Earth, yet it never approaches closer than 55.7 million kilometers, or 140 times the distance of the Moon from the Earth. This makes it a difficult object to observe. A 2- or 3-inch (5- or 7.5-cm) telescope will show whatever polar cap is tilted toward the Earth (assuming the cap is large enough at the time) and a few of the main dark areas, such as Syrtis Major. On the whole, however, I consider a telescope of at least 5 inches (12.5 cm) the minimum necessary for a refractor, and 9 inches (22.5 cm) for a reflector—in the latter case, the mirror must possess a perfect figure, and preferably should be of long focus, say f/9 or f/10.

Most observations tend to be made near the oppositions, which occur at intervals of two years and two months. Since some of the most interesting questions about Mars involve time-dependent changes, however, useful studies can be made, and are strongly encouraged, several months before and after opposition, when the disk is as small as 6″ or 7″ of arc. At the best oppositions, which occur close to the time that Mars passes the perihelion of its orbit, the apparent diameter reaches 25.1″. Unfortunately for Northern Hemisphere observers, the planet is then always low in the sky.[1]

The aphelic oppositions occur in February and March. The disk is much smaller then, of course, only about 14″ of arc, but Mars is higher in the sky—a distinct advantage, since it can be observed through less of the Earth's atmosphere. Also, the Martian atmosphere is then generally clearer.

Indeed, as Schiaparelli pointed out long ago, the size of the disk is less important than the transparency of the Martian atmosphere in determining the visibility of minor features. The clarity of the atmosphere, in turn, depends on the season. The Martian dust storm season generally begins around late spring or early summer in the Martian southern hemisphere. At the perihelic oppositions there may be considerable amounts of dust suspended in the air above Mars, which tends to make the markings appear washed out; at the aphelic oppositions the atmosphere is nearly dust-free (though cirrus clouds are frequent), and in general the contrasts of the markings are much stronger.

Anyone seriously interested in observing the red planet will sooner or later wish to make a permanent record of what is seen. Most observers still carry out their work visually, and this means sketching the planet. A standard scale of 50 mm to the diameter of the planet has been adopted by the British Astronomical Association Mars Section; the American Association of Lunar and Planetary Observers Mars recorders, for some reason, use a scale of only 42 mm. These scales should be adopted in all drawings submitted for the section reports, but in terms of actual work at the telescope, this tends (except near opposition) to make for drawings on the large side. Harold Hill has pointed out that "for a disk of 18 arcseconds, a scale of 50mm corresponds to some 210 inches (5.3 m) to the Moon's diameter for lunar drawings, and for a disk of 9 arcseconds, double that amount! No one, *but no one*, would consider the feasibility of adopting such a scale for lunar work."[2] Hill uses a sliding scale of 3 mm to the arc-second to give a more realistic idea of how the planet looks on a good night. Although the phase of Mars can be ignored close to opposition, at other times it can be quite considerable—at maximum phase, Mars is only 89 percent illuminated and appears as gibbous as the Moon some three or four days from full. Last, I must emphasize that there is no point to drawing the planet unless the seeing is at least reasonably good.

In drawing the planet, it is generally best to begin with a line sketch showing the main reference points—the polar caps and hoods if present, and the outlines of the most prominent features. Once this is finished, one can fill in the finer features in more leisurely fashion. Clouds are conveniently indicated by dashed lines. Because of the planet's rotation, the positions of features are, of course, constantly changing—in general it is best to finish a sketch in fifteen or twenty minutes. One should do one's best to give a realistic portrayal of the markings—too many published drawings show them with a hardness and boldness that is quite misleading; it is difficult to estimate the amount of mischief that has been done by such misrepresentations over the years!

Once mastery of representing Mars in pencil tones has been achieved, one may wish to tackle the colors of the planet. To some extent, the apparent colors are illusory—most notably the bluish tones that sometimes appear in the dark areas, which are produced by simultaneous contrast with the salmon pink deserts. Richard Baum suggests the following technique:

> Take your prepared disk. First lay down a background color, in this case orange-red. Do this by simply scraping off from a pastel (not the oil kind) a certain amount of dust directly on to the center of the disk. Smooth this in with cottonwool (not with finger because of its grease content) and work outwards towards the limb. You will at this stage have a reddish hued disk brighter at the limb, giving a good representation of the limb haze. Lay down the outlines of the markings to be sketched in (very gently as to leave no indentations, but don't use a soft B, rather HB). Then very gently shade in the dark areas, again as to leave no heavy marks. Smooth this detail out by rubbing with the reddish-coated cottonwool, and then gradually work up the shadings into what you require, all the time working gently but consistently—don't hurry the job. The insertion of cloud detail is easily accomplished by the use of a kneadable rubber. Also this type of rubber is very good in that when kneaded to a point, stippling effects on a dusky background are very easy. . . . I originally learned this technique at the age of nine from a superb marine artist, who did his work on clouds and waves and indeed ships in this way with truly superlative results.[3]

I have one of Baum's beautiful and artistic representations of the planet framed on the wall of my study. He gives a rather romantic representation of the planet using a warm orange-red color for the disk. Hill, by contrast, describes his impression of the Martian background as always a pale pink occasionally tinged with ruddiness, adding that "sometimes the warmth can be quite absent." He notes that "the dark markings at times show distinctly blue as shown by Lowell—improbable though such a color may seem. . . . The old adage applies that 'everyone sees in his own way' and especially when confronted with a telescopic image displaying such exquisite delicacy of coloring and shading as does Mars under the best conditions."[4] This is undoubtedly true. The color effects are sensitive to seeing conditions, the telescope's aperture, the disk size, and Martian seasonal effects. During the southern hemisphere summer (L_s = 270–360°), when there is often a great deal of dust in the atmosphere, the contrast of the markings is, as noted earlier, more subdued; the desert areas are then apt to appear more yellowish or even lemon, while the dark areas appear neutral gray or brownish. At the aphelic oppositions, when dust is generally absent, the apparent bluish tints can be rather striking. Various subjective effects also play a role in what one sees, including differences in the response curves of pigment-sensing proteins, or visual pigments, from one individual to the next; the extreme case is complete insensitivity of one or more pigments, also known as color-blindness—remember Schiaparelli!

The discussion of the Martian colors brings us to the next topic —the use of colored filters. Undoubtedly there is no other planet for which their use is more indispensable, and the serious Martian observer simply cannot afford to do without them. A yellow filter (Wratten 12 or 15) increases the contrast of the dark areas with the background—this is what Schiaparelli used, and it always brought the markings out "like spots of India ink." Orange (W21 and W23A) and red (W25) also will increase the contrast of details and assist in the identification of dust clouds, of which more presently. On the other hand, green (W58), blue (W44A), and blue-violet (W47) filters will bring out limb hazes and terminator clouds and frost patches. Since some of these filters are quite dense, they require telescopes with fairly large light grasp—the blue-violet filter (W47) needs at least 9 inches (22.5 cm). Generally speaking, surface details are invisible in the blue and blue-violet filters, but one should be on the lookout for the so-called blue clearings.

Photographically, Mars has always been a difficult subject, and detailed knowledge of photographic emulsions and techniques is required to do it justice. Even state-of-the-art high-resolution film requires exposures of several seconds, which is long enough to hopelessly blur details. In any case, photography has now virtually

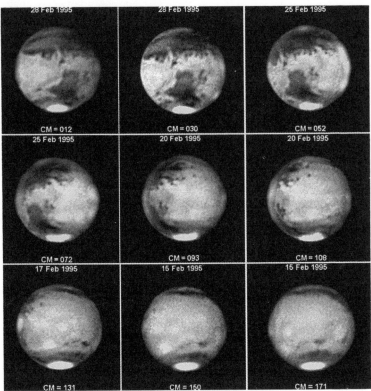

28 Feb 1995	28 Feb 1995	25 Feb 1995
CM = 012	CM = 030	CM = 052
25 Feb 1995	20 Feb 1995	20 Feb 1995
CM = 072	CM = 093	CM = 108
17 Feb 1995	15 Feb 1995	15 Feb 1995
CM = 131	CM = 150	CM = 171

FIGURE 21. Rotation of Mars, 1995. These CCD images were obtained by Donald C. Parker with a Lynxx CCD camera and 16-inch Newtonian between January 28 and March 18, 1995 (the date of opposition was February 12). They have been arranged to correspond with a single rotation of the planet. CM refers to the longitude of the central meridian, and south is at the top.

Among the readily identifiable features, CM = 012 shows Meridiani Sinus (Dawes forked bay) just left of center, followed by the tip of the wedge-shaped Margaritifer Sinus. The large trapezoidal area extending above the north polar cap is Mare Acidalium. CM = 093 shows Solis Lacus

been supplanted by the charge-coupled device, or CCD, so I will say no more about it here.

Still-frame CCD cameras are expensive and require a computer, but in the hands of such pioneers as Jean Dragesco, Isao Miyazaki, and Donald C. Parker they have yielded awe-inspiring images of Mars (fig. 21). Parker's best results to date were obtained in February–March 1995, at an aphelic opposition when the appar-

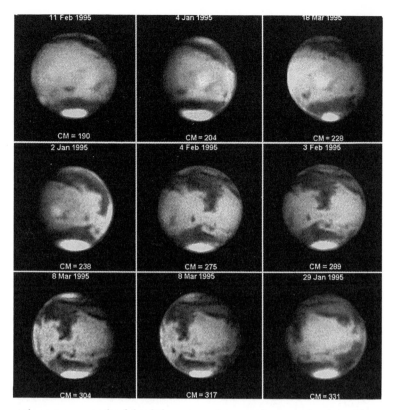

11 Feb 1995	4 Jan 1995	18 Mar 1995
CM = 190	CM = 204	CM = 228
2 Jan 1995	4 Feb 1995	3 Feb 1995
CM = 238	CM = 275	CM = 289
8 Mar 1995	8 Mar 1995	29 Jan 1995
CM = 304	CM = 317	CM = 331

at the extreme south of the disk. CM = 108, 131, and 150 are dominated by the bright Tharsis region. In CM = 150, Mare Sirenum is the dark strip visible at top of disk. CM = 204 features the bright Elysium basin. CM = 238 shows the dark areas Mare Cimmerium and Mare Tyrrhenum, followed by Syrtis Major, which is coming around the limb at far right. CM = 289 is centered on Syrtis Major. Note the tri-lobed Deltoton Sinus along its right upper margin and the ribbon-like Sinus Sabaeus curving to the extreme right. Many of the smaller features are subject to time-dependent changes owing to the effects of windblown dust, as a careful comparison with figure 22 will show. (Courtesy Donald C. Parker)

ent diameter of Mars was never more than 13.8″. He uses a 16-inch (41-cm) Newtonian at his observatory at Coral Gables, Florida, and enjoys extraordinary seeing owing to the generally laminar airflow off the ocean—on many nights, planetary images resemble steel engravings, rippled only now and then by an atmospheric tremor.

In addition to still-frame CCD cameras, video cameras using CCDs sensitive enough for planetary imaging have become commercially available and are quite affordable. As with ordinary photography, one projects the image through an eyepiece, but because of the remarkable resolution of the CCD (as measured by the number of picture elements, or pixels), very high magnifications can be used. Typically, video cameras have exposure times of 1/60 second and recording rates of thirty frames per second, and by playing back the videotape frame by frame, one can follow the moment-to-moment changes in seeing. Though the individual frames are generally far from good, every now and then a few frames will appear sharp and stationary.

Seeing is not the only cause of distortion; another is granularity due to electronic noise. This can be handled in various ways. For example, one can use a "frame grabber" device with image-processing software to construct a final image with an improved signal-to-noise ratio, though such equipment is expensive. A cheaper alternative is simply to composite an image by taking pictures off the monitor using a 35-mm single-reflex camera. Since the signal-to-noise ratio varies as the square root of the number of images, by combining four to eight consecutive frames when the seeing is good, one can produce an image in which the granularity due to electronic noise is reduced by a factor of two or three.[5] However, even the best still frames fall short of capturing what Thomas Dobbins has aptly described as the "ineffable sense of reality" of viewing a videotape.[6]

Many amateurs are now producing CCD images that show detail beyond the reach of even the best visual observers using much larger instruments. Though the trained human eye was never seriously challenged by photography using silver-grain emulsions (requiring exposures of one second or more), its long reign in glimpsing fine planetary details is finally over, surpassed by the CCD.

THE LARGER surface features of Mars are generally stable, and an observer equipped with one of the maps made by Schiaparelli or Flammarion, or even by Beer and Mädler, will easily recognize the main features of the planet. But the maps, even allowing for inevitable errors (and ignoring the canals!), are not identical with modern ones.

This is hardly surprising; the fact of time-dependent changes on Mars is well established. First of all, there are seasonal cycles, which affect the intensity and visibility of the various markings, the most obvious effect being the alternate waxing and waning of the two polar caps. The solid north polar cap is hidden during its deposition phase by a cloudy hood that covers it during much of the northern hemisphere autumn and winter. The cap emerges from the polar hood at the start of spring, at which time it extends to approximately 65° N. In general, the retreat of the north cap is quite symmetrical. In late spring, the cap becomes fissured into two portions by a dark rift, Rima Tenuis. Around the time of the summer solstice (L_s = 90°), the bright mass known as Olympia breaks off, separated from the main cap by the dark rift Rima Borealis. In most years the seasonal carbon dioxide frost cap evaporates off completely, leaving a residual water ice remnant. As it retreats, the north polar cap appears to be surrounded with a dark collar, sometimes known as the Lowell band, which was once regarded as a shallow sea but coincides in position with a wide swath of sand dunes.

The south polar cap is tilted toward the Earth at the perihelic oppositions, and thus is well presented as it rapidly shrinks during the southern spring and summer. It begins to break up by southern hemisphere mid-spring; the most notable remnant is the Mountains of Mitchel, also known as Novissima Thyle, located near the cap's retreating edge (and between Martian longitudes 300° and 330° w). It begins to detach from the polar cap at around L_s = 215°, and is fully separated by L_s = 230°. There are also various dark rifts in the cap, such as the Rima Australis and Rima Angusta. The seasonal carbon dioxide frost cap generally fails to disappear completely — thus, unlike the northern cap, the residual cap consists mostly of carbon dioxide frost rather than water, and is always much smaller than its northern counterpart.

During the northern hemisphere spring and summer (or, equivalently, the southern hemisphere fall and winter), there is generally little dust in the Martian atmosphere, although whitish clouds are frequent at the limb and terminator. The great basins of Hellas, Argyre, and Elysium are usually frost covered and often appear brilliant white, and there are also many smaller patches of frost.

The largest dust storms occur in the southern hemisphere spring and summer. However, the period in which they are known to occur is not confined to a narrow band, and planet-encircling and global storms have begun at points along Mars's orbit ranging from L_s = 204° to 310°. The sites most often associated with initiation

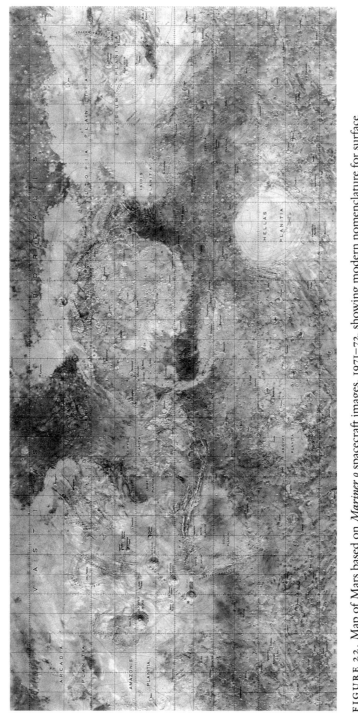

FIGURE 22. Map of Mars based on *Mariner 9* spacecraft images, 1971–72, showing modern nomenclature for surface features. (Map I-961, Miscellaneous Investigation Series, U.S. Geological Survey, 1976)

of dust storm activity are, in the northern hemisphere, Chryse-Acidalia, Isidis–Syrtis Major, and Cerberus; in the southern hemisphere, Hellas, Noachis-Hellespontus, Argyre, and the Solis, Sinai, and Syria Plani regions. There have been no planet-encircling or global storms since 1982. The much-anticipated global storm of 1988 never developed, though regional storms were observed, and there were also regional storms in 1990 and 1992. It now appears that planet-encircling storms (as in 1956, 1971, 1973, and 1977) are the exception, not the rule, but further observations are needed to establish their frequency.

The Martian winds scatter dust around, and the changing dust distribution produces changes in albedo features. E. M. Antoniadi claimed that Syrtis Major underwent regular seasonal changes; its eastern side became streaked and narrow in the spring, then widened in autumn. Nepenthes-Thoth, a conspicuous marking earlier in the century, has now all but disappeared, though Alcyonius Nodus remains conspicuous. There have also been marked changes in Solis Lacus; it was large and complex in the 1970s, but in the 1990s has become smaller and more circular, its classic form up until the 1920s.

THE OBSERVER who wishes to become familiar with the main aspects of the planet as seen in a modest telescope may find the following guide of some use (also, see fig. 22). In addition to the main telescopic markings (albedo features), I highlight aspects of the topography of each Martian region, which, though not directly visible, are convenient to keep in mind when orienting oneself to the disk.

The Martian longitudes begin at 0° in the 0.5-kilometer-wide crater Airy-0 in Sinus Meridiani, and increase progressively to the west; thus Solis Lacus lies at about 90° W longitude; Mare Sirenum is at 180°; and Isidis, the bright area just east of Syrtis Major, is at 270° W. Syrtis Major itself is centered on about 290° W longitude. The sense of rotation is the same as that of increasing longitude; that is, toward the west (right, on the inverted telescope image) at a rate of 14.62°/hour. Because the Martian day is about forty minutes longer than ours, as one watches from night to night the markings appear to fall gradually backward on the disk by some nine degrees each night. The planet thus appears to complete an illusory backward rotation over a period of forty days, during which the entire circumference passes in review before the observer.

It is convenient to begin our tour with the Sinus Sabaeus (Terra Sabaea), since this is the region through which the 0° meridian passes. The serpentine ribbon of Sinus Sabaeus runs just south of the equator and terminates in the Sinus Meridiani (Terra Meridiani), whose two northward-pointing forks are still sometimes referred to as Dawes' forked bay. The point between the forks, christened Fastigium Aryn by Schiaparelli, is the zero of Martian longitudes—or, more precisely, the small crater in this position known as Airy-0 is. The forked appearance is not always apparent, but at times it can be distinct in only a 6-inch (15-cm) telescope.

South of Sinus Sabaeus are the moderately bright regions of Deucalion and Noachis. Their ancient and heavily cratered terrain dates back to the middle Noachian period of heavy bombardment, four billion years ago. The brighter equatorial continent to the north is also heavily cratered and is known as Arabia; among its leading craters are Schiaparelli, which lies just on the border between Sinus Sabaeus and Arabia—it was often seen as a circular brightish patch by E. M. Antoniadi—and Cassini. Again, though, none of these features can be seen with modest telescopes.

To the west of Sinus Sabaeus lies Margaritifer Sinus (Margaritifer Terra), whose beaklike extension sometimes appears broken off at the end. The Ares Valles, one of the largest Martian outflow channels, courses through the region on its way to Xanthe Terra to the northwest; Xanthe Terra is also the site of the great Tiu, Simud, and Shalbatana channels, in which spacecraft photographs have shown lemniscate islands and alluvial plains suggestive of massive flooding. These features originate in western Margaritifer Sinus in the rough-and-tumble region known as "chaotic terrain." The flooding that took place here was on a catastrophic scale—much greater than anything ever seen on Earth. With good seeing, one can see even in modest telescopes that this is a region of complex formation.

The dusky region south of Margaritifer Sinus is occupied by Mare Erythraeum, whose boundary is rather ill-defined; however, there is one notable feature: the large circular formation of Argyre, which lies at about latitude 50° s. It is a splendid feature, 1,500 kilometers across, and was formed by a huge impact late in the era of heavy bombardment. The impact was so violent that the debris fell in several concentric rings; the innermost ring is very rugged and forms the basin's rim, of which the northern part is known as the Nereidum Montes; the southern is known as the Charitum Montes. The basin's floor tends to be covered with frost in the southern hemisphere winter, which causes it to appear brilliant at times.

North of Xanthe Terra is the plain of Chryse Terra, in the middle of which lies the *Viking 1* landing site; still farther north is Acidalia Planitia (Mare Acidalium), one of the vast northern plains. Mare Acidalium is among the most prominent features during the aphelic oppositions, when the northern hemisphere is tilted toward the Earth. One sometimes forgets the scale of what is being unfolded in the telescope; Mars is a small world, but since it has no oceans, its land area is equal to that of the Earth, and Mare Acidalium covers an area equal to about a quarter of the continental United States. In recent years, the extension of Mare Acidalium, Nilokeras, has been prominent enough to be made out easily in small telescopes.

AS MARS rotates, Sinus Meridiani passes off the disk and the Solis Lacus region comes into view. Solis Lacus is a variable albedo feature centered within the bright circular region that Schiaparelli called Thaumasia Felix—the Land of Wonders. It was clearly visible in the 1830s, when Beer and Mädler drew it as small and round; by the 1860s it had become noticeably elongated in an east-west direction. In 1877, Schiaparelli and Trouvelot found it nearly round but slightly elongated north and south. Its east-west elongation was again evident in 1879, and this was the way it generally was figured (though with minor changes) until 1926, when it underwent a radical transformation. In that year Antoniadi found that it curved toward the northwest, at a right angle to its usual direction. In 1939 it was found to be made up of a number of small spots, and its form remained large and complex through the 1970s. In the 1990s, however, it has become smaller and rounder again. Obviously Solis Lacus is one of the most variable regions on the planet, which is hardly surprising given that the plain located here, Solis Planum, is one of the areas long associated with the initiation of dust storm activity.

The great Valles Marineris canyon system runs through the region just south of the equator, its system of interconnected canyons running east and west from Margaritifer Chaos to the complex Noctis Labyrinthus. The Martian canyons are on a stupendous scale compared with the Grand Canyon of the Colorado River; at their widest point, in Melas canyon, the span reaches a width of 200 kilometers. Thus, because of the curvature of the planet, if one stood on the north rim of the canyon, the walls of the south canyon would be completely below the horizon! Even in a small telescope, the course of Valles Marineris can be followed as a curving, dark, threadlike line; this was known as Agathadaemon on the canal-filled maps of the classical era.

South and west of the Valles Marineris complex are the great volcanic plains associated with the huge Tharsis bulge. In a small telescope, this enormous region extending from the edge of Mare Sirenum to the north pole appears bland and featureless, but since the spacecraft explorations it has become one of the most famous regions on Mars, for here lie the great shield volcanoes. Arsia Mons, Pavonis Mons, and Ascraeus Mons run along a southwest-to-northeast line, and Olympus Mons, the tallest mountain in the solar system, is located at longitude 130° w and latitude 20° n. In modest telescopes, one can sometimes make out faint dusky patches in these locations, or, more commonly, whitish patches—the latter, of course, was the form Schiaparelli saw when he discovered Olympus Mons long ago and christened it Nix Olympica (Snows of Olympus). The volcanoes are often overhung by clouds—use blue filter! The shield of Olympus Mons, though it rises some 25 kilometers above the surrounding plains, is 800 kilometers across at the base, so that the slope is not very steep—only about 6°. Thus, immense as Olympus is, its shadow at the terminator is not within the reach of Earth-based observers.

WE CONTINUE to follow the features that come into view as Mars rotates. What Schiaparelli called the "great diaphragm" of the southern hemisphere begins with the strip of Mare Sirenum, which projects eastward toward Solis Lacus; this swath of darkness broadens as it continues on through Mare Tyrrhenum and breaks into a complex of smaller patches—Antoniadi's "leopard skin." The broken or mottled appearance is partly controlled by the underlying relief and indicates the action of wind depositing and sweeping away materials of different colors. The landforms consist of the rough, cratered terrain that occupies so much of the southern hemisphere of Mars; but these topographical features can only be inferred by the telescopic observer—they are nowhere directly observable. Between Mare Sirenum and Mare Tyrrhenum is a lighter-albedo band, Hesperia Planum, which precedes the large, roughly rectangular darkish patch of Mare Tyrrhenum onto the disk. The latter ends in the northward-pointing wedge of Syrtis Minor. Schiaparelli, inspired by the maritime view of Mars, thought that Hesperia was a floodplain or marsh lying between the two adjacent seas, a reasonable supposition at the time; but the spacecraft photographs have shown that it is a geologically distinct unit, the Hesperian system, which consists of ridged plains overlying the older cratered terrain of Noachian age.

The northern hemisphere in this part of the planet is dominated by the bright plains of Amazonis Planitia and Elysium Planitia. The latter is the site of the great Elysium volcanoes—Albor Tholus, Elysium Mons, and Hecates Tholus, of which nothing, of course, can be made out in modest telescopes. Nevertheless, it is always worth looking with the "mind's eye" and recalling that the volcanoes here are inferior only to those of Tharsis itself! One can make out a dark patch, Trivium Charontis–Cerberus, which has been rather faint in recent years. The Elysium basin often appears as a brightish patch and is sometimes frost covered. Though the region has been largely inundated by volcanoes, part of the basin's rim still stands above the volcanic plains and forms a mountain range, the Phlegra Montes, which was identified in the spacecraft photographs.

At aphelic oppositions one can make out Utopia Planitia—landing site of the *Viking 2* spacecraft—in the extreme north, and the dark plains of Vastitas Borealis. The region between 75° and 85° N is peppered with sand dunes.

WE COME finally to the most celebrated area of Mars: Syrtis Major Planitia. The prominent, dark Syrtis Major is clearly shown in a 1659 drawing by Christiaan Huygens. A low-relief shield volcano has been identified within Syrtis Major whose eruptions were the source of the dark materials covering the region; and indeed, the whole region is an elevated volcanic plateau. The southern part of Syrtis Major is streaked and mottled, owing to wind action, and there are large dune fields within its great expanse. In the southwest, close to where the Sinus Sabaeus branches off, lies the great crater Huygens—it is partly filled by dark material and can sometimes be glimpsed from Earth. The peculiar Deltoton Sinus consists of three arcuate, semicircular "bays," or such Antoniadi thought them when he first saw them with the great Meudon refractor in 1909. There are two huge basins in the Syrtis Major region of the planet: Isidis Planitia, which encroaches on Syrtis Major from the northeast, and Hellas, which lies directly to the south and is by far the most prominent basin on Mars. Hellas is 2,100 kilometers across and is enclosed on the east by the darkish strip of Mare Hadriaticum (Hadriaca Patera), and on the west by that of Hellespontus. In winter the Hellas basin is often partly or wholly filled with frost. There are some elusive and evidently variable albedo markings; Schiaparelli sketched crisscrossing canals there, the Peneus and Alpheus; and in 1892, J. M. Schaeberle and Stanley Williams figured a prominent dark patch near the basin's center, which Antoniadi later named Zea Lacus. This has returned to prominence in recent years—it was very marked at the 1988 opposition.

WE HAVE now followed Mars through a complete rotation, and we return once more to our starting point, Sinus Sabaeus. The Martian features have been described in the order in which they appear in the true rotation of the planet. In fact, however, since it is not possible to follow Mars through a complete rotation in a single night, and since the early evening hours are often the most convenient for viewing, in practical terms the observer tends to pursue the slow apparent drift of the markings in backward order from night to night—thus the Sinus Sabaeus gives way to Syrtis Major, followed successively by Mare Tyrrhenum, Mare Sirenum, Solis Lacus, and Margaritifer Sinus.

I must emphasize that what I have given here is only a first sketch of Mars. The novice observer will see little; but with experience, more and more comes into view. Mars is a difficult object to observe, but there is none more rewarding; and the observer's interest is always piqued by the ever-changing panorama of polar caps, dark markings, and clouds and dust storms.

Epilogue

Twenty years after Viking we are finally going back, with no fewer than ten missions planned for the period between 1996 and 2003. In November 1996, the American *Mars Pathfinder* (MPF) is scheduled for launch. If all goes well, it will land in July 1997 at the Ares–Tiu Valles outflow channels (19.5° N, 32.8° W) and will deploy a small robotic rover called Sojourner to explore the Ares Vallis floodplain—a site, incidentally, very close to that originally planned for the *Viking 1* landing in 1976.

Mars Global Surveyor (MGS), an orbiter scheduled for launch at the same time, will arrive in Martian orbit in August 1997 and will begin mapping the planet in January 1998. A Russian orbiter and two landers are also due to set out in fall 1997.

It is likely that human explorers will follow eventually—perhaps as soon as the middle of the twenty-first century. If and when they do, they will owe much to the Mars of romance, to Schiaparelli, Lowell, Wells, Burroughs, and the rest. We have only now begun to awaken from our dream of Mars, the fire opal we have so long sought through our telescopes, to see it as it really is. The warm hues of the Martian deserts prove to be as deceptive as the pure polar pink of an Arctic sunset, and belie the really terrible reality. Their apparent warmth is only that of our imaginations, which they have so long fired. Still, with its great volcanoes, canyons, and dry riverbeds, Mars remains a fascinating, even if a lifeless, world. And its exploration has only just begun.

Appendix 1: Oppositions of Mars, 1901–2035

The table below gives the opposition date, the planet's position in the sky in terms of its right ascension (RA; hour angle from the first point of Aries) and its declination north or south of the equator (the latter is particularly important because, for northern observatories, a far southerly declination interferes with observation of the planet, which must then be viewed through a longer path of the Earth's atmosphere); the apparent size of the disk in seconds of arc; and the distance of the planet in astronomical units (1 a.u. = Earth-Sun distance). Owing to the slight inclination of Mars's orbit to that of the Earth, the minimum separation between the two bodies can actually occur a few days before or after the opposition date.

Opposition date	RA	Declination	Disk (seconds of arc)	Distance (a.u.)
1901 Feb 22	10h 26m	+14° 32′	13.8″	0.678
1903 Mar 29	12h 32m	−00° 05′	14.8″	0.640
1905 May 8	15h 00m	−16° 57′	17.3″	0.543
1907 July 6	19h 01m	−27° 59′	22.7″	0.411
1909 Sept 24	00h 10m	−04° 13′	23.8″	0.392
1911 Nov 25	03h 58m	+21° 43′	18.0″	0.517
1914 Jan 5	07h 05m	+26° 33′	15.1″	0.625
1916 Feb 10	09h 36m	+19° 08′	14.0″	0.675
1918 Mar 15	11h 44m	+05° 55′	14.0″	0.662
1920 Apr 21	13h 57m	−10° 21′	15.8″	0.588
1922 June 10	17h 11m	−25° 55′	20.1″	0.462
1924 Aug 23	22h 19m	−17° 40′	25.1″	0.373
1926 Nov 4	02h 36m	+14° 26′	20.2″	0.465
1928 Dec 21	05h 58m	+26° 39′	15.8″	0.589
1931 Jan 27	08h 42m	+22° 54′	14.0″	0.663
1933 Mar 1	10h 59m	+11° 26′	14.0″	0.675
1935 Apr 6	13h 03m	−03° 52′	15.1″	0.624
1937 May 19	15h 43m	−20° 39′	18.0″	0.515
1939 July 23	20h 13m	−26° 24′	24.1″	0.389
1941 Oct 10	01h 07m	+03° 29′	22.7″	0.414
1943 Dec 5	04h 44m	+24° 24′	17.3″	0.545
1946 Jan 14	07h 44m	+25° 35′	14.8″	0.641
1948 Feb 17	10h 07m	+16° 25′	13.8″	0.678

Opposition date	RA	Declination	Disk (seconds of arc)	Distance (a.u.)
1950 Mar 23	12h 13m	+02° 20′	14.4″	0.651
1952 May 1	14h 34m	−14° 17′	16.6″	0.564
1954 June 24	18h 12m	−27° 41′	21.6″	0.433
1956 Sept 10	23h 26m	−10° 07′	24.8″	0.379
1958 Nov 16	03h 25m	+19° 08′	19.1″	0.494
1960 Dec 30	06h 39m	+26° 49′	15.5″	0.610
1963 Feb 4	09h 15m	+20° 42′	14.0″	0.671
1965 Mar 9	11h 25m	+08° 08′	14.0″	0.669
1967 Apr 15	13h 35m	−07° 43′	15.5″	0.605
1969 May 31	16h 32m	−23° 56′	19.4″	0.486
1971 Aug 10	21h 27m	−22° 15′	24.8″	0.376
1973 Oct 25	02h 00m	+10° 17′	21.2″	0.441
1975 Dec 15	05h 29m	+26° 02′	16.2″	0.570
1978 Jan 21	08h 20m	+24° 06′	14.4″	0.654
1980 Feb 25	10h 37m	+13° 27′	13.8″	0.677
1982 Mar 31	12h 43m	−01° 21′	14.8″	0.637
1984 May 11	15h 13m	−18° 05′	17.3″	0.537
1986 Jul 10	19h 20m	−27° 44′	23.0″	0.406
1988 Sep 28	00h 27m	−02° 06′	23.8″	0.396
1990 Nov 27	04h 13m	+22° 28′	18.0″	0.523
1993 Jan 7	07h 19m	+26° 16′	14.8″	0.628
1995 Feb 12	09h 47m	+18° 11′	13.8″	0.676
1997 Mar 17	11h 54m	+04° 41′	14.0″	0.661
1999 Apr 24	14h 09m	−11° 37′	16.2″	0.583
2001 Jun 13	17h 28m	−26° 30′	20.5″	0.456
2003 Aug 28	22h 38m	−15° 48′	25.1″	0.373
2005 Nov 7	02h 51m	+15° 53′	19.8″	0.470
2007 Dec 28	06h 12m	+26° 46′	15.5″	0.600
2010 Jan 29	08h 54m	+22° 09′	14.0″	0.664
2012 Mar 3	11h 52m	+10° 17′	14.0″	0.674
2014 Apr 8	13h 14m	−05° 08′	15.1″	0.621
2016 May 22	15h 58m	−21° 39′	18.4″	0.509
2018 Jul 27	20h 33m	−25° 30′	24.1″	0.386
2020 Oct 13	01h 22m	+05° 26′	22.3″	0.419
2022 Dec 8	04h 59m	+25° 00′	16.9″	0.550
2025 Jan 16	07h 56m	+25° 07′	14.4″	0.643
2027 Feb 19	10h 18m	+15° 23′	13.8″	0.678

Opposition date	RA	Declination	Disk (seconds of arc)	Distance (a.u.)
2029 Mar 25	12h 23m	+01° 04′	14.4″	0.649
2031 May 4	14h 46m	−15° 29′	16.9″	0.559
2033 Jun 27	18h 30m	−27° 50′	22.0″	0.427
2035 Sept 15	23h 43m	−08° 01′	24.5″	0.382

Appendix 2:
Perihelic Oppositions of Mars, 1608–2035

Opposition date	Distance (a.u.)
Aug. 3, 1608	0.376
Sept. 22, 1625	0.397
Aug. 21, 1640	0.373
July 21, 1655	0.385
Sept. 8, 1672	0.382
Aug. 8, 1687	0.374
Sept. 26, 1704	0.400
Aug. 27, 1719	0.374
July 26, 1734	0.382
Sept. 14, 1751	0.385
Aug. 13, 1766	0.373
July 12, 1781	0.397
Aug. 30, 1798	0.375
July 31, 1813	0.380
Sept. 19, 1830	0.388
Aug. 18, 1845	0.373
July 17, 1860	0.393
Sept. 5, 1877	0.377
Aug. 4, 1892	0.378
Sept. 24, 1909	0.392
Aug. 23, 1924	0.373
July 23, 1939	0.390
Sept. 10, 1956	0.379
Aug. 10, 1971	0.376
Sept. 28, 1988	0.396
Aug. 28, 2003	0.373
July 27, 2018	0.386
Sept. 15, 2035	0.382

Appendix 3: Table of Data for the Planet Mars

Orbital data

Semimajor axis	227,940,000 km (1.52366 a.u.)
Eccentricity	0.0934
Inclination	1.8504°
Longitude of ascending node	49.59°
Longitude of perihelion	335.94°
Mean orbital velocity	24.13 km/sec
Mean synodic period	779.94 days
Mean sidereal motion	0.5204°/day

Physical elements

Axial inclination	25.19°
Length of sidereal day	24 hr, 37 min, 22.66 sec
Mean orbital period	686.98 Earth days
	669.60 Martian solar days
Diameter	6,779.84 kilometers
Polar compression	0.006
Surface area (Earth = 1)	0.2825
Volume (Earth = 1)	0.1504
Mass (Earth = 1)	0.1074
Density (water = 1)	3.93
Mean escape velocity	5.027 km/sec

Source: H. H. Kieffer, B. M. Jakosky, C. W. Snyder, and M. S. Matthews, eds., *Mars* (Tucson: University of Arizona Press, 1993).

Appendix 4: The Satellites of Mars

	Phobos	Deimos
Mean distance from Mars	9,378 km	23,459 km
Sidereal period	7 hr, 39 min, 13.84 sec	30 hr, 17 min, 54.87 sec
Eccentricity	0.0152	0.0002
Inclination	1.03°	1.83°
Diameter	13.3 × 11.1 × 9.3 km	7.6 × 6.2 × 5.4 km

Source: H. H. Kieffer, B. M. Jakosky, C. W. Snyder, and M. S. Matthews, eds., *Mars* (Tucson: University of Arizona Press, 1993).

Notes

CHAPTER I. MOTIONS OF MARS

1. It was Georg Joachim Rheticus, the pupil of Copernicus, according to Kepler, *Astronomia Nova,* in Johannes Kepler, *Gesammelte Werke,* 22 vols., ed. W. von Dyck and Max Caspar (Munich: C. H. Beck, 1937), vol. 3, p. 8.

2. Copernicus, "Commentariolus," in *Three Copernican Treatises,* 2d ed., trans. Edward Rosen (New York: Dover, 1959), pp. 77–78.

3. For biographical details about Tycho Brahe, see Victor E. Thoren, *The Lord of Uraniborg: A Biography of Tycho Brahe* (Cambridge: Cambridge University Press, 1990).

4. Kepler, *Astronomia Nova,* cap. 7, in *Gesammelte Werke,* vol. 3, p. 108. On Kepler, the standard source is Max Caspar, *Kepler,* trans. C. Doris Hellman (1959; reprint, New York: Dover, 1993). Technical accounts of Kepler's calculations are in J. L. E. Dreyer, *A History of Astronomy from Thales to Kepler* (1906; reprint, New York: Dover, 1953); Antonie Pannekoek, *A History of Astronomy* (1961; reprint, New York: Dover, 1989); Alexandre Koyré, *The Astronomical Revolution,* trans. R. E. W. Maddison (1973; reprint, New York: Dover, 1992); and, especially, Curtis Wilson, "How Did Kepler Discover His First Two Laws?" *Scientific American* 226 (1972): 93–106.

5. Kepler, *Astronomia Nova,* cap. 7, in *Gesammelte Werke,* vol. 3, p. 108.

6. The full title is *Astronomia Nova* αἰτιολο ητος *sue Physica Coelestis, tradita Commentariis de Motibus Stellae Martis. Ex Observationibus G. V. Tychonis Brahe.* An English translation is available as *New Astronomy,* trans. William H. Donohue (Cambridge: Cambridge University Press, 1993).

7. Kepler, *Astronomia Nova,* Epistola Dedicatoria, in *Gesammelte Werke,* vol. 3, p. 8; Koyré, *The Astronomical Revolution,* pp. 277–278.

CHAPTER 2. PIONEERS

1. A very good recent English translation has been produced by Albert van Helden, *Sidereus Nuncius, or the Sidereal Messenger* (Chicago: University of Chicago Press, 1989).

2. Galileo Galilei, *Le Opere di Galileo Galilei,* Edizione Nationale, 20 vols., ed. Antonio Favaro (Florence: G. Barbera, 1890–1909; reprinted 1929–39, 1964–66), vol. 10, p. 503.

3. Kepler, *Conversation with Galileo's Sidereal Messenger,* trans. Edward Rosen (New York: Johnson Reprint Co., 1965), pp. 14, 77.

4. Kepler, *Dioptrice* (Augsburg, 1611); trans. in E. S. Carlos, *A part of the preface of Kepler's* Dioptrics, *forming a continuation of Galileo's* Sidereal Messenger (London, 1880; reprint, Dawsons of Pall Mall, 1960), p. 88.

5. Galileo, *Opere,* vol. 10, p. 474.

6. F. Fontana, *Novae Coelestium Terrestriumque rerum observationes* (Naples, 1646).

7. C. Flammarion, *La planète Mars, et ses conditions d'habitabilité,* 2 vols. (Paris: Gauthier-Villars et Fils, 1892), vol. 1, p. vii.

8. See Albert van Helden, "'Angulo Cingitur': The Solution of the Problem of Saturn," *Journal for the History of Astronomy* 5 (1974): 155–174.

9. Huygens's sketches and notes are reproduced in F. Terby, "Aréographie, ou étude comparative des observations faites sur l'aspect physique de la planète Mars depuis Fontana (1636) jusqu'à nos jours (1873)," in *Mémoirs des savants étrangers de l'Académie Royale des Sciences de Belgique* 39 (1875), at p. 8.

10. Flammarion, *La planète Mars,* vol. 1, p. 16.

11. G. D. Cassini to M. Petit, June 18, 1667, "Extrait d'une lettre de M. Cassini . . . a M. Petit," *Le Journal des Sçavans* pour l'anné mdclxii (reprint, Paris, 1729), pp. 122–125.

12. G. D. Cassini, *Martis circa proprium axem revolubilis observationes Bononiae habitae* (Bologna, 1666), and *Dissertatio apologetica de maculis Jovis et Martis* (Bologna, 1666).

13. Robert Hooke, *Philosophical Transactions of the Royal Society of London* (London, 1665–1666), vol. 1, p. 239.

14. C. Huygens, *Oeuvres Complètes,* 21 vols. (The Hague: Société Hollandaise des Sciences, 1888–1950), vol. 21, p. 224.

15. Quoted in Willy Ley, *Watchers of the Skies* (New York: Viking Press, 1963), p. 169.

16. For a complete discussion, see Albert van Helden, *Measuring the Universe: Cosmic Dimensions from Aristarchus to Halley* (Chicago: University of Chicago Press, 1989).

17. Bernard le Bovier de Fontenelle, *Conversations on the Plurality of Worlds,* trans. H. A. Hargreaves (Berkeley: University of California Press, 1990), p. 52.

18. Steven J. Dick, *Plurality of Worlds: The Origins of the Extraterrestrial Life Debate from Democritus to Kant* (Cambridge: Cambridge University Press, 1982), p. 129.

19. Huygens, *Oeuvres,* vol. 21, p. 701; the English edition is *The Celestial Worlds Discover'd: or, Conjectures concerning the Inhabitants, Plants and Productions of the Worlds in the Planets* (London, 1698), p. 21.

20. Quoted in Joseph Ashbrook, *The Astronomical Scrapbook: Skywatchers, Pioneers and Seekers in Astronomy* (Cambridge, Mass.: Sky Publishing, 1984), p. 127.

1. The 1704 observations are described in G. Maraldi, "Observations des taches de Mars pour vérifier sa révolution autour de son axe," *Histoire et Mémoires de l'Académie des Sciences* (Paris, 1706), p. 74; those of 1719 are in "Nouvelle observations de Mars," *Histoire et Mémoires* (Paris, 1720), p. 144. Incidentally, at the 1719 opposition, the planet's brightness, or perhaps its color—it seems to have been even more than usually red that year—caused a panic. Many believed it to be a nova or red comet portending calamity.

2. W. Herschel, "Astronomical Observations on the Rotation of the Planets," *Philosophical Transactions of the Royal Society* 81, pt. 1 (1781): 115.

3. Mrs. J. Herschel, *Memoir and Correspondence of Caroline Herschel* (New York: D. Appleton, 1876), p. 53.

4. W. Herschel, "On the remarkable Appearances at the Polar Regions of the Planet Mars, the Inclination of its Axis, the Position of its Poles, and its spheroidical Figure; with a few Hints relating to its real Diameter and Atmosphere," *Philosophical Transactions of the Royal Society of London* 74 (1784): 233–273.

5. Ibid.

6. Ibid.

7. Ibid.

8. Schroeter's bibliography includes *Beobachtungen vershiedener schwarz-dunkler Flecken des Jupiters* (Lilienthal, 1786); *Beobachtungen über die Sonnefackeln und Sonnenflecken* (Erfurt, 1789); *Selenotopographische Fragmente über den Mond* (Helmstedt, 1791); *Aphroditographische Fragmente zur genauern Kenntniss des Planeten Venus* (Helmstedt, 1796); *Selenotopographische Fragmente über den Mond,* 2d Band (Göttingen, 1802); *Kronographische Fragmente zur genauern Kenntniss des Planeten Saturn* (Göttingen, 1808), a second part was destroyed in the fire of 1813, see below; *Beobachtungen des grossen cometen von 1807 in physischer Hinsicht* (Göttingen, 1811); and *Hermographische Fragmente zur genauern Kenntniss des Planeten Mercur* (Göttingen, 1816).

9. J. H. Schroeter, *Areographische Beiträge zur genauern Kenntnis und Beurtheilung des Planeten Mars,* ed. H. G. van de Sande Bakhuyzen (Leiden, 1881), p. 1.

10. Ibid., p. 2.

11. Though Schroeter confirmed the basic rotation period of the planet, he found irregularities in his measures of its markings which he attributed to real cloud movements. If his measures yielded a period that was too short, he concluded that the markings had an independent motion due to winds blowing in the same direction as the rotation; if too long, the motion was produced by winds blowing in the opposite direction. From such evidence he compiled a table of Martian wind velocities, which, we now know, was sheer fantasy.

12. Ibid., p. 430.

13. Ibid., p. 440.

14. A description of the destruction of Lilienthal is in Richard Baum, "The Lilienthal Tragedy," *Journal of the British Astronomical Association* 101 (1991): 369–371. Baum gives as his source a letter by W. F. Denning to *The Observatory* dated July 4, 1904, while Denning in turn cites J. Hemingway, *The Northern Campaigns and History of the War, from the Invasion of Russia in 1812* (Manchester, 1815).

15. Tischbein has sometimes been accused of being a clumsy engraver and responsible at least in part for the charge often leveled against Schroeter that he was a clumsy draftsman. A comparison of Schroeter's pencil drawings of Mars with Tischbein's copperplates, however, exonerates the engraver.

16. F. Terby, "Aréographie, ou étude comparative des observations faites sur l'aspect physiquede la planète Mars depuis Fontana (1636) jusqu'à nos jours (1873)," *Mémoires des savants étrangers de l'Académie Royale des Sciences de Belgique* 39 (1875).

17. Bakhuyzen, Foreword to Schroeter, *Areographische Beiträge*.

18. Ashbrook, *The Astronomical Scrapbook: Skywatchers, Pioneers and Seekers in Astronomy* (Cambridge, Mass.: Sky Publishing, 1984), p. 290.

CHAPTER 4. AREOGRAPHERS

1. Henry C. King, *The History of the Telescope* (1955; reprint, New York: Dover, 1979), p. 68.

2. Ibid., p. 160.

3. H. de Flaugergues, "Les taches de la planète Mars," *Journal de Physique* (Paris) 69 (1809): 126; Baron Franz Xaver von Zach, *Correspondenz,* vol. 1 (Gotha, 1818), p. 180.

4. See, for example, C. F. Capen and L. J. Martin, "The Developing Stages of the Martian Yellow Storm of 1971," *Lowell Observatory Bulletin,* no. 157, vol. 3, pp. 211–216.

5. G. K. F. Kunowsky, "Einige physische Beobachtungen des Mondes, des Saturns, und Mars," *Astronomisches Jahrbuch für 1825* (Berlin, 1822).

6. C. Flammarion, *La planète Mars, et ses conditions d'habitabilité,* 2 vols. (Paris: Gauthier-Villars et Fils, 1892), vol. 1, p. 101.

7. The most complete information on Mädler is in Heino Eelsau and Dieter B. Hermann, *Johann Heinrich Mädler, 1794–1874* (Berlin: Akademie Verlag, 1985).

8. Beer and Mädler's investigations on Mars were published in *Astronomische Nachrichten* in 1831, 1834, 1835, 1838, and 1839. These investigations were later collected into a book, which appeared in both French and German editions: *Fragmente sur les corps celestes du système solaire* (Paris, 1840), and *Beiträge zur physischen Kenntniss der himmlischen Körper im Sonnen-*

systeme (Weimar, 1841). Mädler's 1841 investigations were published in *Astronomische Nachrichten* in 1842.

9. Beer and Mädler, "Physische Beobachtungen des Mars bei seiner Opposition im September 1830," *Astronomische Nachrichten* 191 (1831): 447–456, at p. 448.

10. Ibid., p. 450.

11. F. Arago, *Astronomie populaire* (Paris, 1854–57), vol. 4, p. 136.

12. A. Secchi, *Osservazioni di Marte, fatte durante l'opposizione del 1858. Memorie dell'Osservatorio del Collegio Romano* (Rome, 1859).

13. A. Secchi, *Osservazioni del pianeta Marte. Memorie dell'Osservatorio del Collegio Romano*, n.s., 2 (Rome, 1863).

14. F. Kaiser, "Untersuchungen über den planeten Mars bei dessen oppositionen in der Jahren 1862 und 1864," in *Annalen der Sternwarte in Leiden*, Dritter Band (The Hague, 1872), pp. 1–87.

15. J. Norman Lockyer, "Measures of the Planet Mars in 1862," *Monthly Notices of the Royal Astronomical Society* 32 (1862): 179–190.

16. E. M. Antoniadi, "The Hourglass Sea on Mars," *Knowledge*, July 1, 1897, pp. 169–172, at p. 170.

17. G. V. Schiaparelli, "Osservazioni astronomiche e fisiche sull'asse di rotazione e sulla topographia del pianeta Marte," *Atti della R. Accademia dei Lincei*, Memoria 1, ser. 3, vol. 2 (1877–78), in *Le Opere di G. V. Schiaparelli*, 10 vols. (1930; reprint, New York: Johnson Reprint Co., 1969), vol. 1, p. 63n.

18. E. Liais, *L'espace céleste et la natur tropicale* (Paris, 1865). He was not the first to suggest that there was vegetation on Mars. That suggestion had already been made by the French astronomer J. H. Lambert, who had died in 1777. Lambert had surmised that the vegetation on Mars was reddish, a view later revived by Camille Flammarion.

19. R. A. Proctor, *Half-hours with the Telescope* (London, 1896), p. 203.

20. Flammarion, *La planète Mars*, vol. 1, p. 204.

21. W. R. Dawes, "On the Planet Mars," *Monthly Notices of the Royal Astronomical Society* 25 (1865): 225–268.

22. The map was first published in Proctor, *Half-hours with the Telescope*, plate 6. It appeared, in somewhat different forms, in Proctor's other works: *Other Worlds than Ours* (London, 1870), p. 92; *The Orbs around Us* (London, 1872), frontispiece; *Essays on Astronomy* (London, 1872), p. 61; etc.

23. G. V. Schiaparelli, "Osservazioni astronomiche e fisiche sull'asse di rotazione e sulla topographia del pianeta Marte," *Atti della R. Accademia dei Lincei*, Memoria 1, ser. 3, vol. 2 (1877–78), in *Le Opere di G. V. Schiaparelli*, 10 vols. (1930; reprint, New York: Johnson Reprint Co., 1969), vol. 1, p. 129.

24. P. A. Secchi, *Memorie dell'Osservatorio del Collegio Romano* (Rome, 1858), vol. 1, no. 3, p. 22.

25. Quoted in W. F. Denning, *Telescopic Work for Starlight Evenings* (London, 1891), p. 125.

1. A. Hall, "My Connection with the Harvard Observatory and the Bonds—1857–1862," in Edward Singleton Holden, *Memoirs of William Cranch Bond and of His Son George Phillips Bond* (San Francisco, 1897), pp. 77–78.

2. A. Hall to S. C. Chandler, Jr., March 7, 1904, quoted in Owen Gingerich, "The Satellites of Mars: Prediction and Discovery," *Journal for the History of Astronomy* 1 (1970): 109–115, at p. 113.

3. A. Hall, "The Period of Saturn's Rotation," *Monthly Notices of the Royal Astronomical Society* 38 (1877–78): 209.

4. A. Hall to E. C. Pickering, February 14, 1888, Harvard College Observatory Archives.

5. A. Hall to E. C. Pickering, February 7, 1888, Harvard College Observatory Archives.

6. A. Hall, "The Discovery of the Satellites of Mars," *Monthly Notices of the Royal Astronomical Society* 38 (1877–78): 205–208, as supplemented by Gingerich, "The Satellites of Mars," and Steven J. Dick, "Discovering the Moons of Mars," *Sky & Telescope* 76 (1988): 242–243.

7. A. Hall to Arthur Searle, October 9, 1877, Harvard College Observatory Archives.

8. A. Hall to E. C. Pickering, October 30, 1877, Harvard College Observatory Archives.

9. Richard A. Proctor, "Note from Mr. Proctor," *Sidereal Messenger* 6 (1887): 259–262, at p. 260.

10. A. Hall to S. C. Chandler, Jr., March 7, 1904, quoted in Gingerich, "The Satellites of Mars."

11. Ibid.

12. N. E. Green, "Observations of Mars, at Madeira in Aug. and Sept. 1877," *Memoirs of the Royal Astronomical Society* 40 (1880): 123–140.

13. H. Pratt, "Notes on Mars, 1877," *Monthly Notices of the Royal Astronomical Society* 38 (1877–78): 61–63, at p. 62.

14. Unfortunately, there is as yet no really satisfactory biography of Schiaparelli. Sources I consulted include Piero Bianucci, "Giovanni Virginio Schiaparelli," *L'Astronomia* 6 (1980): 45–48; E. Fergola et al., *All'Astronomo G. V. Schiaparelli-Omaggio—30 Giugno 1860–30 Giugno 1900* (Milan, 1900), which gives a chronology of his life and lists his most important publications; and G. Cossavella, *L'Astronomo Giovanni Schiaparelli* (Turin, 1914), as well as various necrological sources. The standard edition of Schiaparelli's works is *Le Opere di G. V. Schiaparelli* (1930; reprint, New York: Johnson Reprint Co., 1969).

15. Quoted in Piero Bianucci, "Giovanni Virginio Schiaparelli," *L'Astronomia* 6 (1980): 45–48, at p. 45.

16. Hector MacPherson, *Makers of Astronomy* (Oxford, 1893), p. 189.

17. G. V. Schiaparelli, "Intorno al corso ed all'origine probabile delle

Stella Meteoriche; Lettere al P. A. Secchi," *Bulletino Meteorologico dell'Osservatorio del Collegio Romano,* vols. 5 and 6 (1866–67), in *Opere,* vol. 3, pp. 261–316.

18. R. G. Aitken, "The Orbit of β Delphini," *Popular Astronomy* 11 (1903): 30–34.

19. In a dozen years of observation, Schiaparelli failed to see any spots as marked as those of 1877–78. He announced his result in "Considerazioni sul moto rotatorio del pianeta Venere" (1890), in *Opere,* vol. 5, pp. 361–428. In 1895 he published similar observations which he felt "put the final seal of certainty on the rotation of 224.7 days" (Schiaparelli to F. Terby, July 31, 1895, in *Bulletin de l'Académie de Belgique,* August 1895).

20. The solid body of Venus rotates in a retrograde direction, with a period of 243 days. The surface of the planet is, however, hidden by clouds, which sweep around the planet in only 4 days.

21. Schiaparelli, "Sulla rotazione di Mercurio," *Astronomische Nachrichten,* no. 2944 (1889): 247.

22. Schiaparelli, "Sulla rotazione e sulla costituzione del pianeta Mercurio" (1889), in *Opere,* vol. 5, p. 343.

23. As noted in A. Dollfus and H. Camichel, "La rotation et la cartographie de la planète Mercure," *Icarus* 8 (1968): 216–226.

24. This is not the case in the Southern Hemisphere; if Schiaparelli and Antoniadi had made their observations there, the true rotation period of Mercury might have been established long ago. See B. A. Smith and E. J. Reese, "Mercury's Rotation Period: Photographic Confirmation," *Science* 162 (1968): 1275–1277.

25. Clark R. Chapman and Dale P. Cruikshank, "Mercury's Rotation and Visual Observations," *Sky & Telescope* 34 (1967): 24–26, at p. 25.

26. E. M. Antoniadi, *The Planet Mars,* trans. Patrick Moore (Shaldon Devon, U.K.: Keith Reid, 1975), p. 260.

27. Schiaparelli, "Osservazioni astronomiche e fisiche sull'asse di rotazione e sulla topografia del pianeta Marte," *Atti della R. Accademia dei Lincei,* Memoria 1, ser. 3, vol. 2, in *Le Opere di G. V. Schiaparelli,* 10 vols. (1930; reprint, New York: Johnson Reprint Co., 1969), vol. 1, pp. 11–175, at pp. 11–12.

28. Flammarion, *La planète Mars,* vol. 1, p. 294.

29. Schiaparelli, "Osservazioni astronomiche e fisiche," p. 61. A valuable monograph on the nomenclature of Mars is Jürgen Blunck, *Mars and Its Satellites: A Detailed Commentary on the Nomenclature* (Hicksville, N.Y.: Exposition Press, 1977).

30. Schiaparelli, "Osservazioni astronomiche e fisiche," p. 61.

31. Ibid.

32. P. Lowell, *Mars* (Boston: Houghton Mifflin, 1895), p. 157.

33. Flammarion, *La planète Mars,* vol. 1, p. 301.

34. Schiaparelli, "Il pianeta Marte ed i moderni telescopi" (1878), in *Opere,* vol. 1, pp. 179–200, at p. 188.

35. Concerning the terms *channel* and *canal*, I quote from the *Oxford English Dictionary* (1971): "The words *canel*, CANNEL, and *chanel*, CHANNEL, from the same Latin source, but immediately from old French, were in much earlier use in Eng[lish]; when canal was introduced it was to some extent used as a synonym of these, but the forms were at length differentiated." The term *canal* is, however, taken to refer specifically to "an artificial watercourse constructed to unite rivers, lakes, or seas, and serve the purposes of inland navigation. (The chief modern sense, which tends to influence all the others.)" Schiaparelli himself never approved of *canal;* see Schiaparelli, *Corrispondenza su Marte,* 2 vols. (Pisa: Domus Galilaeana, 1976), vol. 2, p. 259.

36. Giovanni Cossavella, *L'Astronomo Giovanni Schiaparelli,* p. 10.

37. Schiaparelli, "Osservazioni astronomiche e fisiche," p. 164.

38. Color-blind people perceive any modification of the intensity of light as a change of color, and they are more sensitive to contrast effects than those with normal color vision. Though Schiaparelli generally gave hard and sharp outlines to the dark areas and represented slight differences of shade as hard, sharp lines, he did have perceptions of more delicate structure of these areas. Thus he wrote that part of Libya "had the appearance of a shaggy rug, or if you wish, gave the impression of a crowd of minute pores" (in "Osservazioni astronomiche e fisiche sull'asse di rotatione e sulla topografia del pianeta Marte," *Atti della R. Accademia dei Lincei,* Memoria 3, in *Opere,* vol. 1, p. 472). On p. 479 of the same work he quoted his observation of February 4, 1882: "Tyrrhenum is very beautiful and dark. Ausonia well terminated along it, seeming like froth floating above it, or of something porous." He noted that "an analogous resemblance to foam or a surface full of pores has been offered at times by Iapygia and Libya. A fleecy aspect was seen in Iapygia in 1879. And also in Libya. Perhaps with increasing power of the telescope one could judge if there is something real in such unusual appearances."

39. Lowell, *Mars,* p. 157.

40. Schiaparelli, "Osservazioni astronomiche e fisiche," p. 151.

41. Ibid.

CHAPTER 6. CONFIRMATIONS AND CONTROVERSIES

1. Schiaparelli, "Osservazioni astronomiche e fisiche sull'asse di rotazione e sulla topografia del pianeta Marte," *Atti della R. Accademia dei Lincei,* Memoria 2 (Osservazioni dell'opposizione 1879–80), in *Le Opere di G. V. Schiaparelli,* 10 vols. (1930; reprint, New York: Johnson Reprint Co., 1969), vol. 1, p. 244.

2. Ibid., pp. 334–335.

3. Ibid., p. 386.

4. G. V. Schiaparelli to N. E. Green, October 27, 1879, cited by Green in "Mars and the Schiaparelli Canals," *Observatory* 3 (1879): 252.

5. Ibid.

6. As reported in *The Astronomical Register* 17 (1879): 47–48.

7. T. W. Webb, "Planets of the Season: Mars," *Nature* 34 (1886): 213.

8. "Report of the Meeting of the Association held Dec. 31, 1890," *Journal of the British Astronomical Association* 1 (1890): 112.

9. Schiaparelli, "Osservazioni astronomiche e fisiche," p. 302.

10. Schiaparelli, "Osservazioni sulla topografia del pianeta Marte . . . durante l'opposizione 1881–1882 — Communicazione Preliminare," in *Opere,* vol. 1, pp. 381–388, at p. 385.

11. Ibid., pp. 385–386.

12. Flammarion, *La planète Mars, et ses conditions d'habitabilité,* 2 vols. (Paris: Gauthier-Villars et Fils, 1892), vol. 1, p. 362.

13. Quoted in T. J. J. See, "The Study of Planetary Detail," *Popular Astronomy* 4 (1897): 553.

14. Flammarion, *La planète Mars,* vol. 1, p. 446.

15. Ibid., p. 389.

16. R. A. Proctor, "Maps and Views of Mars," *Scientific American,* supplement, 26 (1888): 10659–10660.

17. Flammarion, *La planète Mars,* vol. 1, p. 412.

18. H. Perrotin, "Observations des canaux de Mars," *Bulletin Société Astronomique de France* 3 (1886): 324–329.

19. Terby, "Physical Observations of Mars," *Astronomy and Astro-Physics* 11 (1892): 479–480.

20. E. H. Gombrich, *Art and Illusion: A Study in the Psychology of Pictorial Representation,* 2d ed. (Princeton, N.J.: Bollingen Press, 1961), p. 204.

21. Perrotin, "Les canaux de Mars. Nouveaux changements observes sur cette planète," *Astronomie* 7 (1888): 213–215.

22. Quoted in Flammarion, *La planète Mars,* vol. 1, p. 510.

23. J. E. Keeler, "First Observations of Saturn with the 36-Inch Refractor of the Lick Observatory," *Sidereal Messenger* 7 (1888): 79–83.

24. *New York Common Advertiser,* November 1888.

25. Flammarion, *La planète Mars,* vol. 1, pp. 426–430.

26. Schiaparelli to O. Struve, July 6, 1878, in *Corrispondenza su Marte,* 2 vols. (Pisa: Domus Galilaeana, 1976), vol. 1, pp. 14–18.

27. Flammarion, *La planète Mars,* vol. 1, p. 425.

28. Ibid.

29. Schiaparelli to Terby, June 12, 1890, in *Corrispondenza su Marte,* vol. 2, pp. 29–31.

30. E. S. Holden, "Note on the Opposition of Mars, 1890," *Publications of the Astronomical Society of the Pacific* 2 (1890): 299–300.

31. E. S. Holden, J. M. Schaeberle, and J. E. Keeler, "White Spots on the Terminator of Mars," *Publications of the Astronomical Society of the Pacific* 2 (1890): 248–249.

32. *San Francisco Chronicle*, June 2, 1895.

33. As quoted in Flammarion, "Idée d'un communication entre les mondes," *L'Astronomie* 10 (1891): 282.

34. R. W. Sinnott, "Mars Mania of Oppositions Past," *Sky & Telescope* 76 (1988): 244–246.

35. As noted in G. E. Hale, "The Aim of the Yerkes Observatory," *Astrophysical Journal* 6 (1897): 310–321, at pp. 320–321.

36. Flammarion, "How I Became an Astronomer," *North American Review* 150 (1890): 100–195, at p. 102. For more details about Flammarion, consult his autobiography, *Mémoires: biographiques et philosophiques d'un astronome* (Paris: E. Flammarion, 1911), which covers the first thirty years of his life; and Philippe de la Cotardière and Patrick Fuentes, *Camille Flammarion* (Paris: Flammarion, 1994).

37. Robert H. Sherard, "Camille Flammarion at Home," *San Francisco Call,* May 28, 1893.

38. Flammarion, *La planète Mars,* vol. 1, p. 515.

39. Ibid.

40. Ibid.

41. Ibid., p. 579.

42. Ibid., pp. 580, 586.

43. Ibid., p. 586.

44. C. A. Young, "Observation of the Red Spot of Jupiter," *Observatory* 8 (1885): 172–174.

45. Quoted in W. F. Denning, *Telescopic Work for Starlight Evenings* (London, 1891), p. 125.

46. G. V. Schiaparelli, "Il pianeta Marte ed i moderni telescopi," *Nuova Antologia,* ser. 2, vol. 9 (Rome, 1878), in *Opere,* vol. 1, pp. 179–200, at p. 197. Compare Isaac Newton in the *Opticks* (1707): "If the Theory of making Telescopes could at length be fully brought into Practice, yet there would be certain Bounds beyond which Telescopes could not perform. For the Air through which we look upon the Stars, is in a perpetual Tremor. . . . The only Remedy is a most serene and quiet Air such as may perhaps be found on the tops of the highest Mountains above the grosser Clouds."

47. Agnes M. Clerke, *A Popular History of Astronomy during the Nineteenth Century,* 3d ed. (London: Adam and Charles Black, 1893), p. 519.

48. E. S. Holden, "The Lowell Observatory in Arizona," *Publications of the Astronomical Society of the Pacific* 6 (1894): 160–169, at p. 165.

49. W. H. Pickering, "Mars," *Astronomy and Astro-Physics* 11 (1892): 668–672.

50. Ibid., p. 669.

51. Ibid., p. 670.

52. Clerke, *Popular History of Astronomy,* p. 344.

53. Ibid.

1. William Graves Hoyt, *Lowell and Mars* (Tucson: University of Arizona Press, 1976), p. 12. Lowell's continuing fascination is attested by the fact that he has been the subject of several biographies. In addition to Hoyt's there is a biography by Lowell's brother, Abbott Lawrence Lowell, *Biography of Percival Lowell* (New York: Macmillan, 1935). Ferris Greenslet's *The Lowells and Their Seven Worlds* (Boston: Houghton Mifflin, 1946), contains much original information about Lowell and his family. Another important source is William Lowell Putnam, *The Explorers of Mars Hill: A Centennial History of Lowell Observatory* (West Kennebunk, Maine: Phoenix, 1994). Hoyt's *Planets X and Pluto* (Tucson: University of Arizona Press, 1980) is a thorough account of Lowell's mathematical and observational quest for a trans-Neptunian planet, which occupied much of his time and energy after the turn of the century. What is likely to be the authoritative biography is now being written by David Strauss, who has already completed an excellent account of the early history of the founding of the Lowell Observatory: "Percival Lowell, W. H. Pickering, and the Founding of the Lowell Observatory," *Annals of Science* 51 (1994): 37–58, which I follow closely here.

2. Greenslet, *The Lowells and Their Seven Worlds*, p. 366.

3. Nathan Appleton, *The Introduction of the Power Loom and the Origin of Lowell* (Boston, 1858), p. 9.

4. Amy Lowell, "Sevenels, Brookline, Mass.," *Touchstone* 7 (1920): 210–218.

5. Quoted in Wrexie Louise Leonard, *Percival Lowell: An Afterglow* (Boston: Badger Press, 1921), p. 25.

6. P. Lowell, "Reply to Newcomb," *Astrophysical Journal* 26 (1907): 131.

7. David Strauss, "'Fireflies Flashing in Unison': Percival Lowell, Edward Morse and the Birth of Planetology," *Journal for the History of Astronomy* 24 (1993): 157–169, at p. 160.

8. A. L. Lowell, *Biography of Percival Lowell*, p. 12.

9. Greenslet, *The Lowells and Their Seven Worlds*, p. 349.

10. David Strauss, "The 'Far East' in the American Mind, 1883–1894: Percival Lowell's Decisive Impact," *Journal of American-East Asian Relations* 2 (1993): 217–241, at p. 225.

11. Greenslet, *The Lowells and Their Seven Worlds*, p. 349.

12. Ibid.

13. *The Soul of the Far East* and *Noto* were prepublished in the *Atlantic Monthly*.

14. P. Lowell, "Noto," *Atlantic Monthly* 67 (1891): 500.

15. Greenslet, *The Lowells and Their Seven Worlds*, p. 355.

16. Flammarion later wrote that Lowell had told him that "his passion for astronomy, in particular for the discoveries he made in the world of

Mars, had been inspired by the publication of this book" (C. Flammarion, "Percival Lowell," *Bulletin Société Astronomique de France* 30 [1916]: 423).

17. P. Lowell to E. C. Pickering, February 14, 1894, E. C. Pickering Papers, Harvard University Archives.

18. S. C. Chandler to E. S. Holden, September 4, 1894, Mary Lea Shane Archives of Lick Observatory.

19. The scale was as follows: 0, diffraction disk and rings confused and enlarged; 2, disk and rings confused but not enlarged; 4, disk defined; no evidence of ring; 6, rings broken but traceable; 8, rings complete but moving; 10, rings motionless.

20. P. Lowell to A. E. Douglass, April 16, 1894, Lowell Observatory Archives.

21. The quotation is from Strauss, "Percival Lowell, W. H. Pickering and the Founding of the Lowell Observatory," p. 37.

22. W. W. Campbell, review of *Mars,* by Percival Lowell, *Publications of the Astronomical Society of the Pacific* 51 (1896): 207.

23. Quoted in A. L. Lowell, *Biography of Percival Lowell,* p. 72.

24. P. Lowell, observing notebooks, Lowell Observatory Archives.

25. P. Lowell, *Mars and Its Canals* (New York: Macmillan, 1906), p. 149.

26. P. Lowell, observing notebooks, Lowell Observatory Archives.

27. W. H. Pickering, "The Seas of Mars," *Astronomy and Astro-Physics* 13 (1894): 553–556.

28. Lowell, "The Polar Snows," *Popular Astronomy* 2 (1894): 52–56.

29. Lowell, "Spring Phenomena," *Popular Astronomy* 2 (1894): 97–100.

30. Ibid., p. 100.

31. Lowell, "The Canals—I," *Popular Astronomy* 2 (1894): 255–261, at p. 261.

32. Lowell, "Oases," *Popular Astronomy* 2 (1895): 343–348, at p. 348.

33. Ray Bradbury, Arthur C. Clarke, Bruce Murray, Carl Sagan, and Walter Sullivan, *Mars and the Mind of Man* (New York: Harper and Row, 1973), p. 11.

34. Lowell, "Mars," *Popular Astronomy* 2 (1894): 1–8.

35. Lowell, "The Canals—I," p. 258.

36. J. E. Keeler to G. E. Hale, December 27, 1894, Yerkes Observatory Archives.

37. Greenslet, *The Lowells and Their Seven Worlds,* p. 366.

38. Quoted in A. L. Lowell, *Biography of Percival Lowell,* p. 93. Flammarion published his own account of their conversation in 1896 in "Recent Observations of Mars," *Scientific American* 74 (February 29, 1896): 133–134.

39. Flammarion, "La planète Mars," *L'Astronomie* 13 (1894): 321–329, at p. 328.

40. Flammarion, *La planète Mars, et ses conditions d'habitabilité,* 2 vols. (Paris: Gauthier-Villars et Fils, 1909), vol. 2, p. 135.

41. Flammarion, "Mars and Its Inhabitants," *North American Review* 162 (May 1896): 546–557.

42. Schiaparelli to F. Terby, November 30, 1896, in *Corrispondenza su Marte,* vol. 2, p. 137.

43. Schiaparelli to F. Terby, January 27, 1895, in *Corrispondenza su Marte,* vol. 2, pp. 166–167.

44. Schiaparelli to O. Struve, May 1, 1898, in *Corrispondenza su Marte,* vol. 2, pp. 278–280.

45. Schiaparelli, "Il pianeta Marte," *Opere,* vol. 2, pp. 47–74; the third and fourth parts of this four-part essay appeared as "The Planet Mars," trans. W. H. Pickering, *Astronomy and Astro-Physics* 13 (1894): 635–640, 714–723.

46. Schiaparelli, "La vita sul pianeta Marte," *Natura ed Arte* 4 (June 1, 1895), in *Opere,* vol. 2, pp. 83–95, at p. 88.

47. Ibid., p. 91.

48. Ibid.

49. His comment to Flammarion was published in a note to Flammarion's translation of "La vita sul pianeta Marte," which was published as "La vie sur la planète Mars" in *Bulletin Société Astronomique de France* 12 (1898): 423–442. To George Comstock of the Washburn Observatory of the University of Wisconsin, he explained in 1897: "I pray you not to take [this paper] too seriously; it is a *lupus ingenii* whose purpose is to show that it is possible to explain some of the mysterious phenomena of Mars without assuming anything very extraordinary or different from what one finds on Earth" (*Corrispondenza su Marte,* vol. 2, p. 235).

50. Quoted in Hector McPherson, "The Problem of Mars," *Popular Astronomy* 29 (1921): 129–137, at p. 133.

51. Schiaparelli, *Corrispondenza su Marte,* vol. 2, p. 192.

52. Schiaparelli, "Observations of the Planet Mars," *Opere,* vol. 2, p. 249.

53. E. M. Antoniadi, "Mars Section, Fifth Interim Report, 1909," *Journal of the British Astronomical Association* 20 (1909): 138.

CHAPTER 8. HOW THE EYE INTERPRETS

1. W. W. Campbell, "The Spectrum of Mars," *Publications of the Astronomical Society of the Pacific* 6 (1894): 228–236.

2. E. E. Barnard, *Nashville Artisan,* August 10, 1883. The standard biography of E. E. Barnard is William Sheehan, *The Immortal Fire Within: The Life and Work of Edward Emerson Barnard* (Cambridge: Cambridge University Press, 1995).

3. Barnard, "Mars: His Moons and His Heavens" (unpublished manuscript in the Vanderbilt University Archives).

4. "E. E. Barnard's Visit with G. V. Schiaparelli in Milan" (manuscript in the Vanderbilt University Archives); it was published by George Van Biesbroeck in *Popular Astronomy* 42 (1934): 553–558.

5. Barnard, observing notebooks, Lick Observatory.

6. Ibid.

7. E. E. Barnard to S. Newcomb, September 11, 1894, Simon Newcomb Papers, Library of Congress.

8. Ibid.

9. Ibid.

10. Noted in "Proceedings of the Meeting of the Royal Astronomical Society, April 14, 1882," *Observatory* 5 (1882): 135–137.

11. E. Walter Maunder, "The Tenuity of the Sun's Surroundings," *Knowledge,* March 1, 1894, pp. 49–50, at p. 49.

12. Ibid.

13. Maunder, "The Canals of Mars," *Knowledge,* November 1, 1894, pp. 249–252, at p. 251.

14. Ibid., p. 252.

15. W. G. Hoyt, *Lowell and Mars* (Tucson: University of Arizona Press, 1976), p. 201.

16. A. L. Lowell, *Biography of Percival Lowell* (New York: Macmillan, 1935), p. 66, quoting from an unpublished paper written by Percival Lowell in 1897 which was intended to be the introduction to the first volume of the observatory's *Annals.*

17. A. E. Douglass, "Notice to the Citizens of Flagstaff," March 29, 1895, Lowell Observatory Archives.

18. Lowell, "Detection of Venus' Rotation Period and Fundamental Physical Features of the Planet's Surface," *Popular Astronomy* 4 (1896): 281–285; "Determination of Rotation Period and Surface Character of the Planet Venus," *Monthly Notices of the Royal Astronomical Society* 57 (1897): 148–149; "The Rotation Period of Venus," *Astronomische Nachrichten,* no. 3406 (1897): 361–364; "Venus in the Light of Recent Discoveries," *Atlantic Monthly* 5 (1897): 327–343.

19. Lowell, *Boston Evening Transcript,* November 28, 1896.

20. Lowell, "Mascari, Cerulli and Schiaparelli on Venus' Rotation Period," *Popular Astronomy* 4 (1897): 389.

21. E. M. Antoniadi, "On the Rotation of Venus," *Journal of the British Astronomical Association* 8 (1897): 46.

22. P. Lowell, "Means, Methods, and Mistakes in the Study of Planetary Evolution" (unpublished manuscript dated April 13, 1905, Lowell Observatory Archives).

23. See Andrew T. Young, "Seeing and Scintillation," *Sky & Telescope* 42 (1971): 139–141, at p. 150.

24. Douglass's most important work had been as Pickering's understudy in Arequipa, in which role he had begun his studies of atmospheric "seeing." See Douglass, "Atmosphere, Telescope, and Observer," *Popular Astronomy* 5 (1897): 64–84; also P. Lowell, "Atmosphere: In Its Effect on Astronomical Research" (lecture text, ca. spring 1897, Lowell Observatory Archives).

25. W. S. Adams to E. E. Barnard, October 29, 1905, Vanderbilt University Archives.

26. Douglass to W. L. Putnam, March 12, 1901, Andrew Ellicott Douglass Papers, Special Collections, University of Arizona Library.

27. Douglass's later career is described in George Ernest Webb, *Tree Rings and Telescopes: The Scientific Career of A. E. Douglass* (Tucson: University of Arizona Press, 1983). In his work with tree rings, Douglass thought that he could see evidence of eleven-year sunspot cycles. This aspect of his work was finally refuted by Valmore LaMarche and Harold Fitts, working at Douglass's Laboratory of Tree-Ring Research, in 1972. John A. Eddy noted in his review of Webb's book that "the ultimate irony of Douglass's later life [was] in succumbing to the same faults of autosuggestion that he had diagnosed in Lowell not that many years before" (*Journal for the History of Astronomy* 17 [1986]: 69–71).

28. Lowell, "The Markings on Venus," *Astronomische Nachrichten*, no. 3823 (1902): 129–132, at p. 130.

29. Lowell, "Venus—1903," *Popular Astronomy* 12 (1904): 184–190, at p. 185.

30. Lowell, "Epitome of Results at the Lowell Observatory April 1913–April 1914," *Lowell Observatory Bulletin*, no. 59 (1914).

31. E. Walter Maunder and J. E. Evans, "Experiments as to the Actuality of the 'Canals' of Mars," *Monthly Notices of the Royal Astronomical Society* 58 (1903): 488–499, at p. 498.

32. "Report of the Meeting of the Association Held on Dec. 30, 1903," *Journal of the British Astronomical Association* 14 (1904): 118.

33. Antoniadi, "Report of Mars Section, 1896," *Memoirs of the British Astronomical Association* 6 (1898): 55–102, at p. 62.

34. Antoniadi, "Report of Mars Section, 1903," *Memoirs of the British Astronomical Association* 16 (1910): 55–104, at p. 58.

35. P. B. Molesworth, manuscript in Royal Astronomical Society Archives.

36. For accounts of Cerulli's career, see Mentore Maggini, "Vincenzo Cerulli," *Memorie de la Societa Astronomica Italiana* 4 (1927): 171–187; and Luigi Prestinenza, "Vincenzo Cerulli e la sua Collis Uraniae," *L'Astronomia* (October 1989): 25–28.

37. V. Cerulli, *Marte nel 1896–97* (Collurania, 1898), p. 105.

38. Schiaparelli, *Corrispondenza su pianeta Marte,* 2 vols. (Pisa: Domus Galilaeana, 1976), vol. 2, p. 307.

39. P. Lowell, *Mars as the Abode of Life* (New York: Macmillan, 1910), pp. 181–182.

40. P. Lowell, *Mars and Its Canals* (New York: Macmillan, 1906), p. 126.

41. Quoted in P. Lowell, *Mars as the Abode of Life,* p. 155. This was less an endorsement than it seemed, however, for Schiaparelli wrote to Antoniadi on August 29, 1909: "Perhaps over a longer or shorter period,

molecular actions develop inside the images, in spite of their being fixed—or even during the fixing—and these actions follow different images of the same object, even when the photographs are taken under conditions and circumstances which are apparently identical, and in the same series. All in all, the problem of the photography of Mars is beset with traps and difficulties" (quoted in E. M. Antoniadi, *The Planet Mars,* trans. Patrick Moore [Shaldon Devon, U.K.: Keith Reid, 1975], p. 262).

42. P. Lowell, *Mars and Its Canals,* p. 277.

43. See W. H. Wesley, "Photographs of Mars," *Observatory* 28 (1905): 314; and H. H. Turner, "From an Oxford Note-book," *Observatory* 28 (1905): 336.

44. A. R. Wallace, *Is Mars Habitable?* (London: Macmillan, 1907).

45. Lowell, "A General Method for Evaluating the Surface-Temperature of the Planets; with Special Reference to the Temperature of Mars," *Philosophical Magazine,* ser. 6, 14 (July 1907): 161–176.

46. Wrote Wallace: "The figure he uses in his calculations for the actual albedo of the earth, 0.75, is also not only improbable, but almost self-contradictory, because the albedo of cloud is 0.72, and that of the great cloud-covered planet, Jupiter, is given by Lowell as 0.75, while Zollner made it only 0.62. Again, Lowell gives Venus an albedo of 0.92, while Zollner made it only 0.50 and Mr. Gore 0.65. This shows the extreme uncertainty of these estimates, while the fact that both Venus and Jupiter are wholly cloud-covered, while we are only half-covered, renders it almost certain that our albedo is far less than Mr. Lowell made it. It is evident that mathematical calculations founded upon such uncertain data cannot yield trustworthy results. But this is by no means the only case in which the data employed in this paper are of uncertain value. . . . It requires a practiced mathematician, and one fully acquainted with the extensive literature of this subject, to examine these various data, and track them through the maze of formulae and figures so as to determine to what extent they affect the final result" (*Is Mars Habitable?* p. 51).

47. Ibid., p. 20.

48. Wrexie Louise Leonard, *Percival Lowell: An Afterglow* (Boston: Badger Press, 1921), p. 25. The lectures were serialized in *Century* magazine, and in 1910 were published in book form as *Mars as the Abode of Life.*

49. Lowell to D. P. Todd, July 26, 1907, Lowell Observatory Archives.

CHAPTER 9. OPPOSITION 1909

1. V. Cerulli, "Polemica Newcomb-Lowell-fotografie lunari," *Rivista di astronomia* 2 (1908): 13–23, at p. 13. My thanks to Signor Luigi Prestinenza for bringing this letter to my attention.

2. G. V. Schiaparelli to E. M. Antoniadi, December 15, 1909, quoted in

E. M. Antoniadi, *La planète Mars, 1659–1929* (Paris: Hermann et Cie, 1930), p. 31.

3. "G. Schiaparelli über die Marstheorie von Svante Arrhenius," *Kosmos* 7 (1910): 303.

4. W. W. Campbell to G. E. Hale, May 11, 1908, Mary Lea Shane Archives of the Lick Observatory.

5. P. Lowell to V. M. Slipher, March 16, 1908, Lowell Observatory Archives.

6. W. W. Campbell to G. E. Hale, May 11, 1908, Mary Lea Shane Archives of the Lick Observatory.

7. W. W. Campbell, "Water Vapor in the Atmosphere of the Planet Mars," *Science* 30 (1909): 474–475.

8. See Donald E. Osterbrock, "To Climb the Highest Mountain: W. W. Campbell's 1909 Mars Expedition to Mount Whitney," *Journal for the History of Astronomy* 20 (1989): 77–97.

9. P. Lowell, "The Canals—I," *Popular Astronomy* 2 (1895): 258.

10. See Donald E. Osterbrock, *Pauper and Prince: Ritchey, Hale and Big American Telescopes* (Tucson: University of Arizona Press, 1993), pp. 89–91.

11. For biographical information on Antoniadi, see Richard McKim, "The Life and Times of E. M. Antoniadi, 1870–1944," *Journal of the British Astronomical Association* 103 (1993): 164–170, 219–227.

12. E. M. Antoniadi, "Report of Mars Section, 1896," *Memoirs of the British Astronomical Association* 5 (1897): 82.

13. Ibid.

14. Ibid., p. 83.

15. Ibid., p. 85.

16. For further information about Brenner's strange personality and career, see Joseph Ashbrook, "The Curious Career of Leo Brenner," *The Astronomical Scrapbook* (Cambridge, Mass.: Sky Publishing, 1984), pp. 103–111; Michael Heim, *Spiridion Gopčević: Leben und Werk* (Wiesbaden: O. Harrassowitz, 1966); and for a more sympathetic view, Martin Stangl, "The Forgotten Legacy of Leo Brenner," *Sky & Telescope* 90 (1995): 100–102.

17. Antoniadi, "Report of Mars Section, 1898–1899," *Memoirs of the British Astronomical Association* 9 (1901): 68.

18. Antoniadi, 1898, quoted in C. Flammarion, *La planète Mars, et ses conditions d'habitabilité,* 2 vols. (Paris: Gauthier-Villars et Fils, 1909), vol. 2, pp. 410–411; and 1902, in John Burnett, "British Studies of Mars: 1877–1914," *Journal of the British Astronomical Association* 89 (1979): 136–145, at p. 139.

19. McKim, "The Life and Times of E. M. Antoniadi," p. 168.

20. Antoniadi, "Report of the Mars Section, 1900–1901," *Memoirs of the British Astronomical Association* 11 (1903): 137.

21. Antoniadi, "Report of Mars Section, 1903," *Memoirs of the British Astronomical Association* 16 (1910): 225–274.

22. E. W. Maunder, "A New Chart of Mars," *Observatory* 26 (1903): 351.

23. Antoniadi, "Note on Mr. Lowell's Drawings of Mars," *Journal of the British Astronomical Association* 20 (1909): 42–43, at p. 43.

24. P. Lowell to E. M. Antoniadi, September 26, 1909, Lowell Observatory Archives.

25. According to the work of A. Kolmogoroff, the energy contained in a turbulent eddy is proportional to the five-thirds power of its linear size. Since the squares of the refractive index fluctuations are proportional to the kinetic energies of the eddies, the refractive index fluctuations are very nearly proportional to the eddy sizes. See Andrew T. Young, "Seeing and Scintillation," *Sky & Telescope* 42 (1971): 139.

26. Antoniadi, "Fourth Interim Report for the Apparition of 1909," *Journal of the British Astronomical Association* 20 (1909): 78–81, at p. 79.

27. Antoniadi, "On the Advantages of Large over Small Telescopes in Revealing Delicate Planetary Detail," *Journal of the British Astronomical Association* 21 (1910): 104–106, at p. 105.

28. Antoniadi to P. Lowell, October 9, 1909, Lowell Observatory Archives.

29. Antoniadi to W. H. Wesley, September 25, 1909, Royal Astronomical Society Archives. In another letter, to Barnard, dated December 11, 1909, he deleted the terrestrial analogy, writing simply: "I have seen a few straight, or even parallel lines, but then I was careful to note that such impressions were fleeting . . . and that, whenever definition was quite satisfactory, all such appearances vanished, the face of the planet looking like that of the Moon" (Vanderbilt University Archives).

30. Antoniadi to P. Lowell, October 6, 1909, Lowell Observatory Archives.

31. P. Lowell to E. M. Antoniadi, November 2, 1909, Lowell Observatory Archives.

32. Antoniadi to P. Lowell, November 15, 1909, Lowell Observatory Archives.

33. Antoniadi later felt the need to defend his priority in having seen what a large telescope could show on Mars, pointing out that his first report on his Meudon observations of September 20 had been published eight days before the famous telegram of Edwin B. Frost, director of the Yerkes Observatory, who had answered a question about the success of the Yerkes refractor in showing the canals by wiring: "Yerkes telescope too powerful for canals." Observations similar to Antoniadi's were made by George Ellery Hale and a group at Mount Wilson (including Lowell's former assistant Douglass) using the 60-inch (1.52-m) reflector diaphragmed to 44 inches (1.12 m) to decipher so much detail on the planet that they found it impossible to draw.

34. Antoniadi, "Fourth Interim Report," *Journal of the British Astronomical Association* 20 (1909): 79.

35. Antoniadi, "Fifth Interim Report," *Journal of the British Astronomical Association* 20 (1909): 137.

36. Antoniadi, "Le retour de la planète Mars," *Bulletin Société Astronomique de France* 40 (1926): 348–349.

37. Antoniadi, "Considerations on the Physical Appearance of the Planet Mars," *Popular Astronomy* 21 (1913): 420.

38. P. Lowell, "Our Solar System," *Popular Astronomy* 24 (1916): 427.

39. Antoniadi, "Second Interim Report on the Observations of 1909," *Journal of the British Astronomical Association* 20 (1909): 28.

40. H. G. Wells, *The War of the Worlds* (New York: Harper, 1898). *A Princess of Mars,* by Edgar Rice Burroughs, was first published in *All-Story Magazine* as a six-part serial, February through July 1912; it appeared in book form in 1917. It was the first of Burroughs's Martian tales, followed by *The Gods of Mars* (1918), *The Warlord of Mars* (1919), *Thuvia, Maid of Mars* (1920), *The Chessmen of Mars* (1922), *The Master Mind of Mars* (1928), *A Fighting Man of Mars* (1931), *Swords of Mars* (1936), *Synthetic Men of Mars* (1940), and *Llana of Gathol* (1948).

CHAPTER 10. THE LINGERING ROMANCE

1. Howard Plotkin, "William H. Pickering in Jamaica: The Founding of Woodlawn and Studies of Mars," *Journal for the History of Astronomy* 24 (1993): 101–122. Pickering died a bitter man in 1938 because he felt that his contributions to astronomy had not been appreciated (Walter H. Haas to William Sheehan, personal correspondence, May 31, 1994).

2. Quoted in Richard McKim, "The Life and Times of E. M. Antoniadi, 1870–1944. Part 2: The Meudon Years," *Journal of the British Astronomical Association* 103, no. 5 (1993): 219–227, at p. 223.

3. E. C. Slipher, *The Photographic Story of Mars* (Flagstaff, Ariz.: Northland Press, 1962).

4. E. M. Antoniadi, *La planète Mars, 1659–1929* (Paris: Hermann et Cie, 1930); it has been translated into English by Patrick Moore as *The Planet Mars* (Shaldon Devon, U.K.: Keith Reid, 1975).

5. McKim, "The Life and Times of E. M. Antoniadi, 1870–1944. Part 2: The Meudon Years."

6. H. P. Klein, N. H. Horowitz, and K. Biemann, "The Search for Extant Life on Mars," in *Mars,* ed. H. H. Kieffer, B. M. Jakosky, C. W. Snyder, and M. S. Matthews (Tucson: University of Arizona Press, 1993), p. 1223.

7. P. Lowell, *Mars as the Abode of Life* (New York: Macmillan, 1908), p. 240.

8. G. de Vaucouleurs, *Physics of the Planet Mars,* trans. Patrick Moore (London: Faber and Faber, 1954).

9. W. S. Adams and T. Dunham, "The B Band of Oxygen in the Spectrum of Mars," *Astrophysical Journal* 79 (1934): 308–316; idem, "Water-Vapor Lines in the Spectrum of Mars," *Publications of the Astronomical Society of the Pacific* 49 (1937): 209–211.

10. G. de Vaucouleurs, *The Planet Mars,* trans. Patrick Moore (London: Faber and Faber, 1950), p. 127.

11. E. M. Antoniadi, *The Planet Mars,* p. 54.

12. Ibid.

13. E. M. Antoniadi, "Mars Report, 1909," *Memoirs of the British Astronomical Association* 20 (1915): 37.

14. Slipher, *The Photographic Story of Mars,* p. 39.

15. Ibid.

16. T. E. Thorpe, "Viking Orbiter Observations of the Mars Opposition Effect," *Icarus* 36 (1978): 204–215.

17. P. Lowell, *Mars and Its Canals,* pp. 39–40.

18. G. de Vaucouleurs, *The Planet Mars,* p. 26.

19. Alfred Russel Wallace, *Is Mars Habitable?* (London: Macmillan, 1907), pp. 34–35.

20. Quoted in Samuel Glasstone, *The Book of Mars* (Washington, D.C.: National Aeronautics and Space Administration, 1968), p. 104.

21. G. Fournier, quoted in de Vaucouleurs, *The Planet Mars,* p. 63.

22. Antoniadi, *The Planet Mars,* p. 27.

23. Ibid., p. 29.

24. E. E. Barnard, observing notebook, Yerkes Observatory Archives.

25. G. P. Kuiper, "Visual Observations of Mars, 1956," *Astrophysical Journal* 125 (1957): 307–317.

26. D. G. Rea, B. T. O'Leary, and W. M. Sinton, "The Origin of the 3.58- and 3.69-Micron Minima in the Infrared Spectra," *Science* 147 (1965): 1286–1288.

27. M. E. Chevreul, *The Principles of Harmony and Contrast of Colours,* 3d ed., trans. Charles Martel (London, 1860).

28. Ibid., p. 11.

29. Dean B. McLaughlin, "Volcanism and Aeolian Deposition on Mars," *Geological Society of America Bulletin* 65 (1954): 715–717; McLaughlin, "Interpretation of Some Martian Features," *Publications of the Astronomical Society of the Pacific* 66 (1954): 161–170; McLaughlin, "Wind Patterns and Volcanoes on Mars," *Observatory* 74 (1954): 166–168; McLaughlin, "Further Notes on Martian Features," *Publications of the Astronomical Society of the Pacific* 66 (1954): 221–229; McLaughlin, "Additional Evidence of Volcanism on Mars," *Bulletin of the American Geological Society* 66 (1955): 769–772; McLaughlin, "Changes on Mars, as Evidence of Wind Deposition and Volcanism," *Astronomical Journal* 60 (1955): 261–270; McLaughlin, "The Volcanic-Aeolian Hypothesis of Martian Features," *Publications of the Astronomical Society of the Pacific* 68 (1956): 211–218; McLaughlin, "A New Theory of Mars," *Michigan Alumnus Quarterly Review* 62 (1956): 301–307.

30. T. Saheki, "Martian Phenomena Suggesting Volcanic Activity," *Sky & Telescope* 14 (1955): 144–146.

31. Interestingly, other flares with associated clouds were noted here by S. Tanabe on November 6, 1958, and by S. Fuikui on November 10, which

led Charles Capen to suspect that the region might be actively volcanic at the present time (Capen, unpublished manuscript). Some of the *Viking 1* orbiter images revealed a cloud that cast a shadow of unusual shape that was consistent with an eruption, and was tentatively identified by Leonard Martin of the Lowell Observatory as a steam vent; it is located at 79° w, 16° s, between Solis Lacus and Tithonius Lacus, near the location where the Japanese observers recorded their flares.

32. G. P. Kuiper, "Note on Dr. McLaughlin's Paper," *Publications of the Astronomical Society of the Pacific* 68 (1956): 304–305.

33. Kuiper, "Visual Observations of Mars, 1956."

34. Quoted in Glasstone, *The Book of Mars,* p. 119.

35. Tombaugh's view is that "the Martian canals were global fracture fault lines, perhaps induced by internal heating and expansion, bursting a thick rigid crust at the round spots known as oases, or triggered by asteroid impacts at the round spots known as oases" (Clyde W. Tombaugh to William Sheehan, personal correspondence, December 6, 1986–March 9, 1987). For the suggestions of Baldwin and Öpik, see Ralph B. Baldwin, *The Face of the Moon* (Chicago: University of Chicago Press, 1949); and E. J. Öpik, "Collision Probabilities with the Planets," *Proceedings of the Royal Irish Academy* 54A (1951): 165–199. Among others who recognized that a flat Mars did not follow from the apparently smooth terminator was V. A. Firsoff, who wrote in *Our Neighbour Worlds* (London: Hutchinson, 1952), p. 23: "On the Moon five-power binoculars will reveal a wealth of detail, but if her phase is full the mountains will not stand out clearly, as they cast no visible shadows. This is how we see Mars at opposition through the most powerful telescopes. . . . Moreover, Mars has a sensible atmosphere, and, although this is remarkably transparent, it does intervene, especially at the limb or terminator, where accidents of the terrain will show best. To sum up, it is perhaps not very significant that no mountains or marked differences in altitude of the ground have been discovered on Mars by direct observation, and our estimates of heights and depressions on his surface must needs be somewhat problematic."

36. A. Dollfus, "Mèsure de la quantité de vapeur d'eau contenue dans l'atmosphere de la planète Mars," *Comptes Rendu Académie Sciences* 256 (1963): 3009–3011; H. Spinrad, G. Münch, and L. D. Kaplan, "The Detection of Water Vapor on Mars," *Astrophysical Journal* 137 (1963): 1319–1321.

37. L. D. Kaplan, G. Münch, and H. Spinrad, "An Analysis of the Spectrum of Mars," *Astrophysical Journal* 139 (1964): 1–15.

CHAPTER 11. SPACECRAFT TO MARS

1. A. J. Kliore, D. L. Cain, G. S. Levy, V. R. Eshleman, G. Fjeldbo, and F. D. Drake, "Occultation Experiment: Results of the First Direct Measurement of Mars' Atmosphere and Ionosphere," *Science* 149 (1965): 1243–1248.

2. R. B. Leighton and B. C. Murray, "Behavior of Carbon Dioxide and Other Volatiles on Mars," *Science* 153 (1966): 136–144. Leighton and Murray calculated that, for a moonlike Mars—that is, one without condensible gases—the temperature would fall well below −128°C in the polar regions, and predicted the existence of permanent caps of frozen carbon dioxide.

3. R. B. Leighton et al., *Mariner Mars 1964 Project Report: Television Experiment*. Part I: *Investigators' Report* (Jet Propulsion Laboratory Technical Report 32–884, part I, 1967).

4. See E. M. Shoemaker, "Impact Mechanics at Meteor Crater," in *The Moon, Meteorites and Comets,* ed. B. M. Middlehurst and G. P. Kuiper (Chicago: University of Chicago Press, 1963), 301–336. For historical background on this particular issue, see W. G. Hoyt, *Coon Mountain Controversies* (Tucson: University of Arizona Press, 1987).

5. See Don E. Wilhelms, *To a Rocky Moon* (Tucson: University of Arizona Press, 1993).

6. J. E. Mellish, letter, *Sky & Telescope* 31 (1966): 339. His letter of January 18, 1935, to Walter Leight describing his observations was published by Rodger W. Gordon, "Mellish and Barnard—They Did See Martian Craters!" *Journal of the Association of Lunar and Planetary Observers* 29 (1975): 196–199.

7. The literature on this subject has grown rather extensive. See, for instance, Rodger W. Gordon, "Martian Craters from Earth," *Icarus* 29 (1976): 153–154; "Craters on Mars and Mercury: A History of Predictions and Observations," in *Yearbook of Astronomy,* ed. P. Moore (London, 1982), pp. 138–155; R. J. McKim, note, *Journal of the British Astronomical Association* 97 (1987): 191–192; William Sheehan and Richard McKim, "The Myth of Earth-Based Martian Crater Sightings," *Journal of the British Astronomical Association* 104 (1994): 281–286; and Richard McKim and William Sheehan, "Visibility of Martian Craters from Earth," *Journal of the British Astronomical Association* 105 (1995): 137.

8. Thomas A. Mutch, Raymond E. Arvidson, James W. Head III, Kenneth L. Jones, and R. Stephen Saunders, *The Geology of Mars* (Princeton: Princeton University Press, 1976), p. 17.

9. D. G. Rea, "Evidence for Life on Mars," *Nature* 200 (1964): 114–116; C. Sagan and J. B. Pollack, "A Windblown Dust Model of Martian Surface Features and Seasonal Changes," *Smithsonian Astrophysical Observatory Special Report* 255 (1967); and Sagan and Pollack, "Windblown Dust on Mars," *Nature* 223 (1969): 791–794.

10. C. F. Capen, *The Mars 1964–1965 Apparition,* Jet Propulsion Laboratory Technical Report 32–990.

11. C. F. Capen to W. Sheehan, personal communication, April 24, 1983.

1. C. R. Chapman, J. B. Pollack, and C. Sagan, *An Analysis of the Mariner 4 Photographs of Mars,* Smithsonian Astrophysical Observatory Special Report 268 (Washington, D.C., 1968).

2. C. F. Capen, "Martian Yellow Clouds—Past and Future," *Sky & Telescope* 41 (1971): 117–120, at p. 120.

3. An intriguing record of what several prominent personalities in planetary astronomy and science fiction expected to find is in Ray Bradbury, Arthur C. Clarke, Bruce Murray, Carl Sagan, and Walter Sullivan, *Mars and the Mind of Man* (New York: Harper and Row, 1973).

4. For detailed discussions, see Thomas A. Mutch, Raymond E. Arvidson, James W. Head III, Kenneth L. Jones, and R. Stephen Saunders, *The Geology of Mars* (Princeton: Princeton University Press, 1976); Michael H. Carr, *The Surface of Mars* (New Haven: Yale University Press, 1981); Victor R. Baker, *The Channels of Mars* (Austin: University of Texas Press, 1982); Peter Cattermole, *Mars: The Story of the Red Planet* (London: Chapman and Hall, 1992); and Hugh H. Kieffer, Bruce M. Jakosky, Conway W. Snyder, and Mildred S. Matthews, eds., *Mars* (Tucson: University of Arizona Press, 1993).

5. J. M. Boyce, *A Method for Measuring Heat Flow in the Martian Crust Using Impact Crater Morphology,* NASA TM-80339 (Washington, D.C., 1979), pp. 114–118.

6. The scale height over which the barometric pressure is reduced by a factor of $1/e$ (where e is the base of natural logarithms, or approx. 2.72), is on the order of 10 kilometers.

7. D. E. Wilhelms and S. W. Squyres, "The Martian Hemispheric Dichotomy May Be Due to a Giant Impact," *Nature* 309 (1984): 138–140.

8. V. R. Baker, *The Channels of Mars* (Austin: University of Texas Press, 1982).

9. V. C. Gulick and V. R. Baker, "Fluvial Valleys and Martian Paleoclimates," *Nature* 341 (1989): 514–516.

10. V. R. Baker and D. J. Milton, "Erosion by Catastrophic Floods on Mars and Earth," *Icarus* 23 (1974): 27–41.

CHAPTER 13. VIKINGS—AND BEYOND

1. Viking Orbiter Imaging Team, *Viking Orbiter Views of Mars,* NASA SP-441 (Washington, D.C., 1980), p. 4.

2. Four spacecraft (two orbiters and two landers) were launched during summer 1973 and arrived at Mars between February 10 and March 12, 1974. Unfortunately, none worked entirely as planned. The braking rocket of the first orbiter, *Mars 4,* failed to fire, and it merely swept past Mars and returned a few television images, none very good. The other orbiter, *Mars 5,*

did make it into orbit and managed to obtain photographs of which a few of the best were comparable in quality to those of *Mariner 9,* but it failed after only twenty-two orbits. As for the landers, the *Mars 7* descent module separated prematurely and missed Mars altogether; *Mars 6* did better, at least initially—the descent module separated successfully and transmitted data all the way down to the surface, but then contact was lost. After that, until the *Phobos* mission of 1989, the Russians concentrated their efforts on Venus instead of Mars, where they enjoyed one of their greatest triumphs in achieving the first successful soft landings on the surface of another planet with their *Veneras 9* and *10,* in October 1975.

3. The true color of the rocks is chocolate brown. See Andrew T. Young, "What Color Is the Solar System?" *Sky & Telescope* 69 (1985): 399–403.

4. Viking Imaging Team, *The Martian Landscape,* NASA SP-425 (Washington, D.C., 1978), p. 36.

5. Not everyone accepted the revised version, even at the time; see, for instance, V. A. Firsoff, *The Solar Planets* (Newton Abbot, U.K.: David and Charles, 1977), pp. 134–135.

6. See, for instance, the summaries provided by H. P. Klein, "The Viking Biological Experiments on Mars," *Icarus* 34 (1978): 666–674, and "The Viking Mission and the Search for Life on Mars," *Review of Geophysics and Space Physics* 17 (1979): 1655–1662; N. H. Horowitz, *To Utopia and Back: The Search for Life in the Solar System* (New York: Freeman, 1986).

7. See G. V. Levin and P. A. Straat, "Recent Results from the Viking Labeled Release Experiment on Mars," *Journal of Geophysical Research* 82 (1977): 4663–4668; Klein, "The Viking Biological Experiments on Mars"; G. V. Levin and P. A. Straat, "A Reappraisal of Life on Mars," in *The NASA Mars Conference,* ed. D. B. Reiber (San Diego: Univelt, Inc., for NASA and the American Astronautical Society, 1988), pp. 187–208.

8. K. L. Jones, S. L. Bragg, S. D. Wall, C. E. Carlston, and D. G. Pidek, "One Mars Year; Viking Lander Imaging Observations," *Science* 204 (1979): 799–806.

9. H. H. Kieffer, S. C. Chase, Jr., T. Z. Martin, E. D. Miner, and F. D. Palluconi, "Martian North Pole Summer Temperatures: Dirty Water Ice," *Science* 194 (1976): 1341–1344.

10. V. A. Firsoff, *The Solar Planets,* p. 141.

11. C. Sagan, J. Veverka, P. Fox, R. Dubisch, J. Lederberg, E. Levinthal, L. Quam, R. Tucker, J. B. Pollack, and B. A. Smith, "Variable Features on Mars: Preliminary Mariner 9 Television Results," *Icarus* 17 (1972): 346–372; see also *Mariner Mars 1971, Project Final Report,* Jet Propulsion Laboratory Technical Report 32-1550 (1973).

12. E. S. Barker, R. A. Schron, A. Woszcyk, R. G. Tull, and S. J. Little, "Mars: Detection of Atmospheric Water Vapor during the Southern Hemisphere Spring and Summer Season," *Science* 170 (1970): 1308–1310.

13. Certainly there are many instances of local obscurations, but noth-

ing like the global storms in the earlier records. During the 1890s and early 1900s, Percival Lowell, who saw only one dust cloud during eleven years of observation, wrote in *Mars and Its Canals* (p. 89): "If we are uncertain of the precise character of the Martian climate, we know on the other hand a good deal about the Martian weather. A pleasing absence of it over much of the planet distinguishes Martian conditions from our own. That we can scan the surface as we do without practical interruption day in and day out proves the weather over it to be permanently fair. In fact a clear sky, except in winter, and in many places even then, is not only the rule, but the rule almost without exceptions. In the early days of Martian study cases of obscuration were recorded from time to time by observers, in which portions of the disk were changed or hidden as if clouds were veiling them from view. More modern observations fail to support this deduction, partly by absence of instances, partly by other explanation of the facts. Certainly the recorded instances are very rare."

14. For obliquities above 54°, the average yearly solar insolation becomes greater at the poles than at the equator (this is actually the case for Uranus, with obliquity 97.9°, and the Pluto-Charon system, with obliquity 118.5°).

15. A. W. Ward, "Climatic Variations on Mars. 1. Astronomical Theory of Insolation," *Journal for Geophysical Research* 84 (1974): 7934–7939.

16. O. B. Toon, J. B. Pollack, W. Ward, J. A. Burns, and K. Bilski, "The Astronomical Theory of Climatic Change on Mars," *Icarus* 44 (1980): 552–607; J. B. Pollack and O. B. Toon, "Quasi-Periodic Climate Changes on Mars: A Review," *Icarus* 50 (1982): 259–287; Bruce M. Cordell, "Mars, Earth, and Ice," *Sky & Telescope* 72 (1986): 17–22.

17. V. R. Baker, R. G. Strom, V. C. Gulick, J. S. Kargel, G. Komatsu, and V. S. Kale, "Ancient Oceans, Ice Sheets and the Hydrological Cycle on Mars," *Nature* 352 (1991): 589–594.

CHAPTER 14. THE HURTLING MOONS OF MARS

1. Jonathan Swift, *Gulliver's Travels* (London, 1726), part 3, chapter 3.

2. See Owen Gingerich, "The Satellites of Mars: Prediction and Discovery," *Journal for the History of Astronomy* 1 (1970): 109–115.

3. B. P. Sharpless, "Secular Accelerations in the Longitudes of the Satellites of Mars," *Astronomical Journal* 51 (1945): 185–195.

4. I. S. Shklovskii and C. Sagan, *Intelligent Life in the Universe* (New York: Dell, 1967).

5. Joseph A. Burns, "Contradictory Clues as to the Origin of the Martian Moons," in *Mars,* ed. H. H. Kieffer, B. M. Jakosky, C. W. Snyder, and M. S. Matthews (Tucson: University of Arizona Press, 1993), 1283–1301.

6. See J. A. Burns, "On the Orbital Evolution and Origin of the Mar-

tian Moons," *Vistas in Astronomy* 22 (1978): 193–210; and, for a particularly lucid and provocative account, Tom Van Flandern, *Dark Matter, Missing Planets and New Comets* (Berkeley, Calif.: North Atlantic Books, 1993).

CHAPTER 15. OBSERVING MARS

1. In this respect, the July 10, 1986, opposition was one of the worst possible, since despite the large apparent diameter (22″ of arc) the planet was almost as far south as it can get—between –23° and –28°—throughout the prime observing window. On the other hand, conditions at the September 28, 1988, opposition were as favorable as they had been at any time since 1909—the apparent diameter was 23″, and the planet lay close to the celestial equator so that an observer at mid-northern latitudes had to peer through only half as much air mass as was necessary in 1986.

2. Harold Hill to William Sheehan, personal correspondence, June 21, 1994.

3. Quoted in Richard McKim, BAA *Mars Section Circular* (London: British Astronomical Association, May 1994).

4. Harold Hill to William Sheehan, personal correspondence, June 21, 1994.

5. Shigemi Numazawa, "Using a CCD on the Planets," *Sky & Telescope* 83 (1992): 209–215.

6. According to amateur Mars observer Thomas Dobbins: "The still images contain anamorphic distortions (i.e., squares deformed into rectangles or trapezoids, circles into ellipses) induced by atmospheric turbulence, despite the extremely short exposures involved. . . . These distortions are further compounded by a poor signal-to-noise ratio." On the other hand, he noted that "when one views video pictures at rates above the threshold for 'flicker fusion,' our eye-brain combination fills in parts of the picture that are missing for short intervals. . . . By filling in or 'integrating' the scenes, we average out noise and smooth out incrementally displaced or distorted images" (Thomas Dobbins to William Sheehan, personal correspondence, February 17, 1995).

Selected Bibliography

Antoniadi, E. M. *La planète Mars, 1659–1929*. Paris: Hermann et Cie, 1930. English translation: *The Planet Mars*. Trans. Patrick Moore. Shaldon, Devon: Keith Reid, 1975.

Baker, V. R. *The Channels of Mars*. Austin: University of Texas Press, 1982.

Batson, R. M., P. M. Bridges, and J. L. Inge. *Atlas of Mars: The 1:5M Map Series*. NASA SP-438. Washington, D.C.: National Aeronautics and Space Administration, 1979.

Blunck, Jürgen. *Mars and Its Satellites*. 2d ed. Hicksville, N.Y.: Exposition Press, 1982.

Bradbury, Ray, Arthur C. Clarke, Bruce Murray, Carl Sagan, and Walter Sullivan. *Mars and the Mind of Man*. New York: Harper and Row, 1973.

Carr, M. H. *The Surface of Mars*. New Haven: Yale University Press, 1981.

Cattermole, Peter. *Mars*. London: Chapman and Hall, 1982.

Cotardière, Philippe de la, and Patrick Fuentes. *Camille Flammarion*. Paris: Flammarion, 1994.

Crowe, Michael J. *The Extraterrestrial Life Debate, 1750–1900: The Idea of a Plurality of Worlds from Kant to Lowell*. Cambridge: Cambridge University Press, 1986.

Firsoff, V. A. *The New Face of Mars*. Hornchurch: Ian Henry, 1980.

Flammarion, Camille. *La planète Mars et ses conditions d'habitabilité*. 2 vols. Paris: Gauthier-Villars et Fils, 1892 and 1909.

Glasstone, Samuel. *The Book of Mars*. NASA SP-179. Washington, D.C.: National Aeronautics and Space Administration, 1968.

Hartmann, W. K., and Odell Raper. *The New Mars: The Discoveries of Mariner 9*. NASA SP-337. Washington, D.C.: National Aeronautics and Space Administration, 1974.

Horowitz, Norman H., *To Utopia and Back: The Search for Life in the Solar System*. New York: Freeman, 1986.

Hoyt, W. G. *Lowell and Mars*. Tucson: University of Arizona Press, 1976.

Kieffer, H. H., B. M. Jakosky, C. W. Snyder, and M. S. Matthews. *Mars*. Tucson: University of Arizona Press, 1992.

Lowell, Percival. *Mars*. Boston: Houghton Mifflin, 1895.

———. *Mars and Its Canals*. New York: Macmillan, 1906.

———. *Mars as the Abode of Life*. New York: Macmillan, 1910.

Moore, Patrick. *Guide to Mars*. New York: W. W. Norton, 1977.

Mutch, T. A., R. E. Arvidson, J. W. Head, K. L. Jones, and R. S. Saunders. *The Geology of Mars*. Princeton: Princeton University Press, 1976.

Putnam, William Lowell. *The Explorers of Mars Hill: A Centennial History of Lowell Observatory, 1894–1994*. West Kennebunk, Maine: Phoenix, 1994.

Richardson, R. S. *Exploring Mars*. New York: McGraw-Hill, 1954.

Richardson, R. S., and Chesley Bonestell. *Mars*. New York: Harcourt, Brace and World, 1964.

Schiaparelli, G. V. "Osservazioni astronomiche e fisiche sull'asse di rotazione e sulla topographia del pianeta Marte." *Atti della R. Academia dei Lincei,* Memoria 1, ser. 3, vol. 2 (1877–78).

Sheehan, W. *Planets and Perception: Telescopic Views and Interpretations, 1609–1909.* Tucson: University of Arizona Press, 1988.

Slipher, E. C. *A Photographic History of Mars.* Flagstaff, Ariz.: Northland Press, 1962.

Vaucouleurs, G. de. *The Planet Mars.* Trans. P. Moore. London: Faber and Faber, 1950.

———. *Physics of the Planet Mars: An Introduction to Areophysics.* Trans. P. Moore. London: Faber and Faber, 1954.

Viking Lander Imaging Team. *The Martian Landscape.* NASA SP-425. Washington, D.C.: National Aeronautics and Space Administration, 1978.

Viking Orbiter Imaging Team. *Viking Orbiter Views of Mars.* NASA SP-441. Washington, D.C.: National Aeronautics and Space Administration, 1980.

Wallace, Alfred Russel. *Is Mars Inhabited?* New York: Macmillan, 1907.

Index

Note: References to figures are in italics.

William Sheehan was born in Minneapolis, Minnesota. He received a B.A. degree from the University of Minnesota, an M.A. degree from the University of Chicago, and an M.D. degree from the University of Minnesota Medical School. He completed a residency program in psychiatry at the University of Minnesota Medical School and now lives in Hutchinson, Minnesota, with his wife, Deborah, and their two children, Brendan and Ryan. Professionally, he is a psychiatrist. He is also an amateur astronomer and historian of science who has published three other books: *Planets and Perception: Telescopic Views and Interpretations, 1609–1909,* which was named book of the year by the Astronomical Society of the Pacific; *Worlds in the Sky: Planetary Discovery from Earliest Times through Viking and Voyager,* an astronomy book club main selection; and *The Immortal Fire Within: The Life and Work of Edward Emerson Barnard,* the authoritative biography of one of the greatest American astronomers of the late nineteenth and early twentieth centuries. He has published more than thirty papers on astronomy and the history of science in journals such as *Astronomy, Sky & Telescope,* the *Journal for the History of Astronomy,* and the *Journal of the British Astronomical Association.*